Efficient Clustering Algorithms for Multivariate Data: A Computational Approach

Integrating Factor Analysis, K-Means, and Validation Measures

Mehmet Cevri

Efficient Clustering Algorithms for Multivariate Data: A Computational Approach

Integrating Factor Analysis, K-Means, and Validation Measures

ELIVA PRESS

ISBN: 978-99993-4-588-0

Cover Design: Eliva Press
Cover Image: www.ingimage.com
Email: info@elivapress.com
Website: www.elivapress.com

This Eliva Press imprint is published by the registered company Eliva Press Global Ltd. part of Eliva Press S.R.L. Publishing Group

The registered company address is: Pope Hennessy Street Level 2, Hennessy Tower Port Louis, Mauritius
Eliva Press S.R.L. Publishing Group legal address is: Bd. Cuza-Voda 1/4 of. 21, Chisinau, Moldova, Europe

Efficient Clustering Algorithms for Multivariate Data: A Computational Approach

ASSOC.PROF. DR. MEHMET CEVRİ

Efficient Clustering Algorithms for Multivariate Data: A Computational Approach

PREFACE

This book focuses on the development of efficient clustering algorithms for large multivariate datasets and presents a computational framework that integrates factor analysis, cluster analysis, and cluster validation techniques. A novel methodology is proposed to identify meaningful structures in high-dimensional data while improving clustering accuracy and computational efficiency. In particular, the silhouette index is employed as a robust validation criterion for determining the optimal number of clusters within the K-means framework.

To evaluate the effectiveness of the proposed algorithm, a comprehensive case study is conducted using a dataset consisting of 40 *Teucrium* species (Lamiaceae), characterized by 21 micromorphological variables. The *Teucrium* dataset serves as a real-world application for assessing the performance of the proposed methodology in handling complex multivariate biological data. Extensive comparative analyses are performed against widely used clustering techniques and validation approaches to demonstrate the strengths and limitations of the proposed framework.

The computational implementation is carried out using Mathematica software, taking advantage of its powerful symbolic and numerical capabilities. Experimental results indicate that the proposed algorithm provides more reliable cluster structures and achieves superior performance in identifying natural groupings within large multivariate datasets. Furthermore, the integration of factor analysis and K-means clustering with silhouette-based clustering validation contributes to improved interpretability and stability of the clustering results.

This book is intended for researchers, data scientists, and students interested in computational statistics and pattern recognition. Overall, the methodology presented in this book offers an effective computational approach for multivariate data clustering and can be applied across diverse scientific fields, including biology, medicine, pharmaceutical research, and other disciplines that require the analysis and classification of complex high-dimensional datasets.

The present book is an expanded and revised version of research previously reported by the author in the article "*An Efficient Algorithm Based on Cluster Analysis for Exploring the Structure of Large Multivariate Datasets*", published in Computers in Biology and Medicine (197, 2025, 111016). The scope of the original study has been considerably broadened through the inclusion of additional theoretical background, methodological developments, simulation results, and discussions that could not be accommodated within the limitations of a journal article.

The remainder of the book is organized into four chapters. Chapter 1 introduces the challenges associated with the analysis of high-dimensional multivariate datasets and motivates the development of validated clustering frameworks. Chapter 2 presents the theoretical foundations of the study, covering Factor Analysis for dimensionality reduction, the K-means clustering algorithm, and two widely used cluster validation measures—the Silhouette Coefficient and Gap Statistic—for assessing clustering quality and determining the optimal number of clusters. Chapter 3 reports the results of extensive simulation studies conducted to evaluate the effectiveness and robustness of the proposed methodology across diverse multivariate data structures. Finally, Chapter 4 discusses the principal findings, highlights the main contributions of the work, and outlines potential directions for future research in multivariate data clustering and validation.

MEHMET CEVRİ

Istanbul, Türkiye

June 2026

ACKNOWLEDGEMENTS

This work was supported by the Scientific Projects Coordination Unit of Istanbul University under Project No: 41895. We also express our sincere gratitude to Assoc. Prof. Dr. Gülay Ecevit Genç from the Department of Pharmaceutical Botany at Istanbul University for kindly providing the Teucrium data employed in the empirical analyses presented in this book.

LIST OF FIGURES

LIST OF TABLES

CONTENTS

CHAPTER 1

INTRODUCTION

Most of real-world problems are complex and have many variables. This makes it challenging to study and properly interpret the dispersion matrix. The number of pairwise correlations between the variables is excessive, rendering further consideration impractical. For an overview of data mining approaches for very large datasets, see [1, 2].

Methods of multivariate data analysis such as cluster and factor analyses help to find relationships among tabulated multivariable and have already been applied in almost all scientific disciplines such as metabolomic data, proteomic and gene microarray, as well as in medical diagnosis [3]. Cluster analysis (CA) is a method of grouping objects based on the information contained within the data, whereby the objects are grouped together on the basis of the similarities identified within the data set. In this context, it is used to divide data into meaningful groups that should reflect the inherent design of the data. It has long been a fundamental tool in a multitude of scientific disciplines, including pattern recognition, statistics, data mining, psychology, machine learning, biology, and other social sciences. For more detailed information about clustering methods, see references [4-6].

Factor Analysis (FA) is a statistical technique used to analyze a dataset in which several intercorrelated quantitative dependent variables are observed. In order to identify patterns among a set of variables, it is possible to apply the technique of data examination in order to ascertain patterns. Alternatively, the number of variables may be reduced to a more feasible level by utilizing a smaller number of index variables, which are known as factors [7]. It is the most prevalent and widely utilized multivariate statistical technique. The genesis of this technique can be dated to the early 1900s, when Charles Spearman developed the Two-Factor Theory [8] to better understand human ability. The objective of factor analysis is to ascertain the fundamental factors that explain the interrelationships among the original variables. This analytical technique is used by scientists in almost all scientific disciplines [9].

The genus Teucrium comprises perennial plants belonging to the Lamiaceae family, which has a wide geographical distribution and comprises 236 genera with over 7000 species [10]. The micromorphological characteristics of several Teucrium species, including nutlet and leaf, have been subjected to comprehensive examination. The systematic values of these characters in the context of specific and infraspecific segregation of the taxa have been definitively determined by Ecevit et al. [11] and Kremer et.al.[12]. A total of 27 species belonging to the genus Teucrium in the Lamiaceae family have been identified in the flora of Turkey and the East Aegean Islands [13]. Nevertheless, the vast majority of these species are primarily concentrated within the Mediterranean phytogeographical region of Turkey. Notwithstanding the description of multiple new species and subspecies in Turkey, the current number of taxa included in 38 species and 8 sections is 50 [14-16].

In this book, we analyzed the classification of Teucrium species based on their micromorphological features such as leave, nutlet and trichome structure. To accomplish this task Factor analysis algorithm is conducted and coded in Mathematica software based on application of the K-means clustering algorithm with the silhouette index. Furthermore, the method is evaluated in comparison to existing algorithms and approaches employed in cluster analysis. This kind of analysis helps obtain some pharmaceutical compounds that are useful in pharmaceutical manufacturing or other industrial fields

CHAPTER 2
MATERIALS AND METHODS

This section presents the foundational materials and statistical methodologies utilized for the classification of complex, multivariate datasets. Initially, factor analysis is introduced to reduce dimensionality and identify the core variables driving species classification; its theoretical underpinnings and practical implementations are elucidated in Section 2.1. Subsequently, cluster analysis techniques are applied to categorize species into homogeneous groups based on their micromorphological similarities, with the general framework detailed in Section 2.2. To provide a comprehensive computational approach, the K-means clustering algorithm is delineated in Section 2.2.1, while the silhouette coefficient method—employed to ascertain the optimal number of clusters and evaluate partitioning performance—is expounded in Section 2.2.2. All computational implementations and graphical visualizations throughout this study were executed using Wolfram Mathematica.

2.1. Factor analysis

It is assumed that *m* represents the number of underlying *common factors* $\{F_i\}^m_{i=1}$, and *p* denotes the number of variables. Let we have a set of variables $\{X_j\}^p_{j=1}$, with $p(\rangle m)$ components, has mean $\{\mu_j\}^p_{j=1}$ and covariance matrix $\boldsymbol{\Sigma}$ of $\{X_j\}^p_{j=1}$. It is designed to reproduce the maximum correlations. Factor analysis represents each of said variables as a linear combination of a smaller set of unobservable factors, in addition to a factor that is unique to each of the response variables. The representation of these linear combinations is given by the following equation:

$$X_j = \mu_j + \sum_{i=1}^{m} l_{ji} F_i + e_j,\ (j = 1,2,...p), \qquad (1)$$

where the error e_j represents a specific factor for variable X_j, while l_{ji} denotes the coefficient of F_i in the linear combination describing X_j. This term is employed to signify the *loading* of the jth variable of the ith factor. The factor loadings provide insight into the extent to which a variable contributes to a given factor. The greater the factor loading, the more a variable contributes to that factor [8]. Factor loadings indicate how strongly a given variable is correlated with a particular factor. This concept is discussed in [17,18]. To ascertain that all variables are equally weighted, it is necessary to standardize them in accordance with the following criteria: each variable must have a value of zero for the mean and a value of one for the variance. This is achieved by standardizing each variable when the variables are measured with different units. In such cases, a global scaling factor is required. Furthermore, when variables are standardized, the covariance matrix is replaced with a correlation matrix. Accordingly, Eq. (1) can be represented in matrix notation as follows:

$$\mathbf{x} = \mathbf{X} - \boldsymbol{\mu} = \mathbf{L}.\mathbf{F} + \mathbf{e}\,, \qquad (2)$$

where $\mathbf{X}$ is $(p \times 1)$ a vector of p observed random variables and mean $E[\mathbf{X}] = \boldsymbol{\mu}$. The vector of *error*, or specific error terms, is denoted by $\mathbf{e}$. $\mathbf{L}$ is $(p \times m)$ matrix whose columns are factor loadings. The linear coefficient $\mathbf{F}$ is $(m \times 1)$ a vector whose components are factors or common factors. The aim of factor analysis is to ascertain a reduced set of distinct summary variables that adequately describe the relationships between the original variables. This is accomplished by eliminating any redundant or duplicated elements within a set of correlated variables. As outlined in Eq. (2), the orthogonal model is predicated on a set of fundamental assumptions, namely:

(i) Factors are orthogonal, so $E(\mathbf{F}) = \mathbf{0}$ and $Cov(\mathbf{F}) = E[\mathbf{F}\mathbf{F}^T] = \mathbf{I}_m$.

(ii) $E[\mathbf{e}] = \mathbf{0}$ and $Cov[\mathbf{e}] = E[\mathbf{e}\mathbf{e}^T] = \boldsymbol{\Psi} = diag(\Psi_1, \Psi_2, \ldots, \Psi_p)$.

(iii) $\mathbf{F}$ and $\mathbf{e}$ are independent, so $Cov(\mathbf{F}, \mathbf{e}) = E[\mathbf{e}\mathbf{F}^T] = \mathbf{0}$.

Furthermore, $\boldsymbol{\Psi}$ is a diagonal matrix comprising the specific variances, which are regarded as a real, positive-definite matrix. If standardized measurements are used, we replace $\mathbf{\Psi}$ with the sample correlation matrix $\mathbf{R}$. Under the orthogonal factor model, covariance matrix $\mathbf{\Sigma}$ can be written as

$$\mathbf{\Sigma} = \mathrm{Cov}(\mathbf{X}) = Var[\mathbf{LF} + \mathbf{e}] = \mathbf{LL}^T + \mathbf{\Psi}. \tag{3}$$

If we use Eq. (3),

$$\begin{aligned} Var(\mathbf{X}_i) &= (\sigma_{ii})^2 = \sum_{j=1}^{m} (l_{ij})^2 + \mathbf{\Psi}_i = h_i^2 + \mathbf{\Psi}_i \\ Cov(X_i, X_j) &= (\sigma_{ij})^2 = \sum_{k=1}^{m} l_{ik} l_{jk},\ i \neq j, \end{aligned} \tag{4}$$

where the first term, $h_i^2 = \sum_{j=1}^{m} l_{ij}^2$, is referred to as *communality*, which is the fraction of the variance of X_i explained common factors $\{F_i\}_{i=1}^{m}$. This percentage value also indicates the reliability of the variable in determining the factors. The second term in Eq. (4), $\mathbf{\Psi}_i$, is called the ith *specific variance* or *uniqueness*. This concept relates to the variability in X_i. The error is attributed to the ith variable, X_i. The use of the standardized version x_j^* of x_j results in $Var(x_j^*) = 1 = h_j^2 + \sigma_j^2$.

The covariance is not dependent on the specific factors in any way. The common factors account for the relationship (covariance/correlation) between the observed variables X_i's

$$Cov(\mathbf{X}, \mathbf{F}) = E\left[(\mathbf{LF} + \mathbf{e})\mathbf{F}^{\mathrm{T}}\right] = \mathbf{L}. \tag{5}$$

That is, $Cov(X_i, F_j) = l_{ij}$. We have p eigenvalues for this variance-covariance matrix $\mathbf{\Sigma}$ as well as corresponding eigenvectors $v_1, v_2, ..., v_p$ for this matrix. By using spectral decomposition of $\mathbf{\Sigma}$, the expression can be rewritten in the following format as a function of the eigenvectors and eigenvalues.

$$\mathbf{\Sigma} = \mathbf{v\Lambda v}^{\mathrm{T}} \cong \sum_{j=1}^{p} \left(\sqrt{\lambda_j}\, v_j\right)\left(\sqrt{\lambda_j}\, v_j\right)^{\mathrm{T}} = \mathbf{L.L}^{\mathrm{T}}, \tag{6}$$

where (λ_j, v_j), $j = 1, 2, ..., p$ are the p eigenvalue-eigenvector pairs of $\mathbf{\Sigma}$ with $\lambda_1 \geq \lambda_2 \geq, ... \geq \lambda_p \geq 0$. This yields the following estimator for the factor loadings:

$$\hat{l}_{ij} = \sqrt{\hat{\lambda}_j}\hat{v}_{ji}. \quad (7)$$

Once the final $p-m$ eigenvalues have reached a sufficiently low value, it is permissible to disregard the contribution of $\sum_{j=m+1}^{p}\left(\sqrt{\lambda_j}\,v_j\right)\left(\sqrt{\lambda_j}\,v_j\right)^{\mathrm{T}}$ to $\mathbf{\Sigma}$. Instead of summing from 1 to *p*, we now sum from 1 to *m* in Eq. (6), ignoring the last *p* - *m* terms in the sum, and obtain reduced value of $\mathbf{\Sigma}$. In this case, specific variances may be taken to be diagonal elements of $\mathbf{\Sigma} - \mathbf{L}.\mathbf{L}^{\mathrm{T}}$. Let $\mathbf{S}$ is the sample covariance matrix, estimated specific variances are provided by diagonal elements to the matrix $S - \hat{\mathbf{L}}.\hat{\mathbf{L}}^T$ so,

$$\hat{\Psi} = \mathrm{diag}(\hat{\Psi}_1, \hat{\Psi}_2, \ldots, \hat{\Psi}_p) \text{ with } \hat{\Psi}_i = s_{ii} - \sum_{j=1}^{m}\hat{l}_{ij}^2, \quad (8)$$

where the third term is taken as the estimated value of the communalities. In order to estimate the factor loadings $\hat{l}_{ij}$ in the factor loadings matrix $\mathbf{L}$, Eq. (7) is used.

Factor scores indicate the estimated values (scores) of each unit in accordance with their common structure. By employing the factor loadings determined in Eq. (7) and using reduced value of $\mathbf{\Sigma}$, factor scores of each variable in accordance with factor structures can be calculated. This is achieved through the utilization of the least squares method. Considering Eq. (2), the vector of common factors for subject *i*, or $\hat{\mathbf{F}}_i$, is determined by minimizing the sum of the squared errors function $E(\mathbf{F})$:

$$E(\mathbf{F}) = \sum_{j=1}^{p}\left(e_{ij}\right)^2 = (\mathbf{x} - \mathbf{LF})^{\mathrm{T}}(\mathbf{x} - \mathbf{LF}). \quad (9)$$

This is like *a least squares regression*, except in this case we already have estimates of the parameters (the factor loadings) but wish to estimate the explanatory common factors. At the minimum of $E(\mathbf{F})$, we have

$$\nabla E = \frac{\partial E}{\partial \mathbf{F}} = (0, \ldots, 0). \quad (10)$$

If we substitute Eq. (9) in Eq. (10), we find the following normal equation

$$\mathbf{L}^T\mathbf{L}\mathbf{F} = \mathbf{L}^T\mathbf{x}. \tag{11}$$

In matrix notation *the least square estimation* of $\mathbf{F}$ is expressed as follows:

$$\hat{\mathbf{F}}_i = \left(\hat{\mathbf{L}}^{\mathrm{T}}\hat{\mathbf{L}}\right)^{-1}\hat{\mathbf{L}}^{\mathrm{T}}\left(\mathbf{X}_i - \overline{\mathbf{X}}\right). \tag{12}$$

In practice, the estimated factor loadings are substituted into this expression, along with the sample mean for the data. Factor scores refer to estimator of $\mathbf{F}$. Using factor loadings in Eq. (7) this yields the *factor scores* as:

$$\hat{\mathbf{F}}_i = \left(\frac{\hat{v}_1^T}{\sqrt{\hat{\lambda}_1}}\left(\mathbf{X}_i - \overline{\mathbf{X}}\right) \quad \frac{\hat{v}_2^T}{\sqrt{\hat{\lambda}_2}}\left(\mathbf{X}_i - \overline{\mathbf{X}}\right) \quad . \;.\; . \quad \frac{\hat{v}_m^T}{\sqrt{\hat{\lambda}_m}}\left(\mathbf{X}_i - \overline{\mathbf{X}}\right) \right)^T, \tag{13}$$

where $\frac{1}{\sqrt{\hat{\lambda}_j}}\hat{v}_j^{\mathrm{T}}$ is called *factor score coefficient* and v_1 through v_m are our first m eigenvectors.

In order to determine the appropriateness of the data for factor analysis, it is possible to employ the Kaiser-Meyer-Olkin (KMO) criterion [19,20] for testing the sufficiency of the sample. The KMO index is calculated using the correlation matrix **R** of the sample data, as follows:

$$\mathrm{KMO} = \frac{\sum_i \sum_j R_{ij}^2}{\sum_i \sum_j R_{ij}^2 + \sum_i \sum_j \rho_{ij}^2} \quad (i \neq j), \tag{14}$$

where R_{ij} represents the correlation between variables, while ρ_{ij} is the partial correlation. Values of the KMO measure in the range of 0.5 to 1.0 indicate that factor analysis is necessary. Accordingly, the sample may be considered adequate if the KMO value exceeds 0.5. Conversely, values less than 0.5 indicate that factor analysis may be unsuitable for this particular context [19, 21].

2.2. Cluster analysis

Use of appropriate measures not only improves the quality of information selection, but also helps reduce the time and processing costs. Therefore, several similarity measures [22] are proposed and used under different applications in statistical literature. The most popular one among them is Euclidian distance which is typically employed to identify the linear distance between vectors, $\mathbf{x}=(x_1,x_2,...,x_p)$ and $\mathbf{y}=(y_1,y_2,...,y_p)$. This is defined as follows:

$$d(\mathbf{x},\mathbf{y})=\left[\sum_{j=1}^{p}\left|x_j-y_j\right|^2\right]^{1/2}. \tag{15}$$

The Euclidean distance metric allows for the clustering of observations with high values of features, as well as observations with low values of features. The distance between two vectors is zero when they are perfectly correlated. Once a distance or similarity measure has been selected, the subsequent step is to ascertain which clustering procedures should be employed. While a hierarchical clustering algorithm was utilized in the analysis, alternative clustering algorithms may be considered for evaluation in order to enhance classification.

2.2.1. The K-means clustering algorithm

Once you have selected the distance or similarity measure, you must choose the appropriate clustering algorithm. Two common clustering methods are partitioning clustering, such as *K-means* clustering, and agglomerative hierarchical clustering. In partitioning clustering, the analyst divides the data sets into a pre-specified number of *K* groups. Thus, the method is referred to as the *K-Means algorithm* [23]. In contrast, hierarchical clustering does not necessitate that the analyst pre-determines the number of clusters to be generated. The K-means clustering algorithm is founded upon the notion of centroids, which are employed to delineate clusters. It is a method that has been extensively employed and subjected to rigorous investigation within the domain of data mining.

Let us consider a data set $D=\{x_i \mid i=1,2,...,m\}$, partitioned into k clusters, $\mathbf{C}=\{C_1,C_2,...,C_k\}$, with $k\le m$. Each cluster C_i comprises a random sample of observations of a d-dimensional Euclidean variable x, with a total of n observations. The aim is to split the data set into exactly m clusters. In order to maintain simplicity throughout, it is assumed that the value of m is already established. The K-means algorithm is applied to assign the set of points to k clusters, with k initial centers selected at random. The sum of squared error (SSE) is given by the following formula

$$J = \sum_{i=1}^{m} \sum_{j=1}^{k} \delta_{ij} \left\| x_i - \mu_j \right\|^2, \tag{16}$$

where μ_j is the mean of cluster $C_j = \{1,2,...,k\}$ and δ_{ij} is the cluster indicator variable with

$$\delta_{ij} = \begin{cases} 1, & x_i \in C_j \\ 0, & \text{otherwise} \end{cases}. \tag{17}$$

We must now minimize J to reach the optimal value. We will differentiate J with respect to two variables, one after the other. In the first step, our objective is to minimize J. This is achieved by treating μ_j as a constant and δ_{ij} as a variable. The resulting values are presented below:

$$\frac{dJ}{d\delta_{ij}} = \sum_{i=1}^{m} \sum_{j=1}^{k} \left\| x_i - \mu_j \right\|^2 = 0, \tag{18}$$

A reduction in this value will result in the following outcome:

$$\delta_{ij} = \begin{cases} 1, & j = \arg\min_j \left\| x_i - \mu_j \right\|^2 \\ 0, & \text{otherwise} \end{cases}. \tag{19}$$

In this step, it can be observed that the data point x_i is assigned to the centroid μ_i, which is the nearest with respect to the Euclidean distance.

In the second step, minimize J by treating δ_{ij} as constant and μ_j as a variable. The following results were obtained:

$$\frac{dJ}{d\mu_j} = 2\sum_{i=1}^{m} \delta_{ij} \left\| x_i - \mu_j \right\| = 0, \tag{20}$$

By reducing this, you will get:

$$\mu_j = \frac{\sum_{j=1}^{k} \delta_{ij} x_i}{\sum_{j=1}^{k} \delta_{ij}}. \tag{21}$$

This step is carried out to recompute the centroid of the cluster after the data points have been assigned. The 1st and 2nd steps will be carried out iteratively until the optimal segregation of data points is achieved. This process separates the objects into groups, allowing us to calculate the metric to be minimized. This process is repeated iteratively until the centroids cease to undergo significant change, or until a predefined stopping condition is met. At this point, the K-means algorithm has successfully grouped the data points into '*K*' clusters, with each data point belonging to the cluster represented by its nearest centroid. It must be emphasized that K-means does not provide an indication of the number of clusters present in the data set. Instead, it minimizes the Euclidean norm.

In order to demonstrate the convergence of the *K*-means algorithm, let $\left\{\mu_1^{(t)}, \mu_2^{(t)}, ..., \mu_k^{(t)}\right\}$ and $\left\{C_1^{(t)}, C_2^{(t)}, ..., C_k^{(t)}\right\}$ denote the means and clusters at the start of the tth iteration of *K*-means. If $\mu_j^{(t+1)} = \mu_j^{(t)}, for\ \forall j = 1,2,3,..,k$, then

$$\begin{aligned} J\left(\mu_1^{(t)}, \mu_2^{(t)}, ..., \mu_k^{(t)}\right) &= \sum_{i=1}^{m}\sum_{j=1}^{k} \delta_{ij} \left\|x_i - \mu_j^{(t)}\right\|^2 \geq \sum_{i=1}^{m} \delta_{ij} \min \left\|x_i - \mu_j^{(t+1)}\right\|^2 \\ &\geq \sum_{i=1}^{m}\sum_{j=1}^{k} \delta_{ij} \left\|x_i - \mu_j^{(t+1)}\right\|^2 = J\left(\mu_1^{(t+1)}, \mu_2^{(t+1)}, ..., \mu_k^{(t+1)}\right). \end{aligned} \tag{22}$$

As the *K*-means algorithm progresses, the sum of the squared error function monotonically decreases, indicating that the algorithm converges in a finite number of iterations.

2.2.2. *Silhouette Coefficient*

The identification of the optimal number of clusters in cluster analysis is a fundamental and challenging problem that requires meticulous consideration. Consequently, a variety of methodologies have been proposed [24, 25]. One of the most frequently employed techniques is the utilization of silhouette coefficients [24, 26]. The performance evaluation of the silhouette index is provided by Starczewski and Krzyżak [27]. Furthermore, it has been proven that the utilization of silhouette coefficients for clustering enables the generation of compact clusters [28]. Their experimental results confirm very good performances of the index.

It is of the greatest importance to ascertain whether the number of clusters obtained is accurate. This is particularly important for algorithms that require this value as a parameter. The silhouette index is an effective tool for this purpose. So as to produce silhouette coefficients, two essential elements are required: firstly, the partition generated through the application of a clustering technique; and secondly, the comprehensive set of inter-object proximities. In this analysis, we consider $\mathbf{X}=\left\{X_1,X_2,...,X_p\right\}$ to be the data set, $\mathbf{C}=\left(C_1,C_2,...,C_k,...,C_K\right)$ to be its clustering into K clusters, where C_k indicates the kth cluster, with $k=1,2,...,K$. For each object X_i assigned to the cluster C_k, let $a\left(X_i\right)$ defined as the mean distance between X_i and every the other objects X_k belonging to the same cluster C_k. The value of $a\left(X_i\right)$ is therefore an indication of the degree of correct assignment of x_i to its designated cluster, C_k. The calculation of this value is conducted via the following formula:

$$a\left(X_i\right)=\frac{\sum_{X_k\in C_k} d\left(X_i,X_k\right)}{\left|C_k\right|-1}, \tag{23}$$

where $\left|C_k\right|$ represents the number of objects within cluster C_k.

The average dissimilarity between X_i and other objects X_l belonging to the cluster C_l, where $l=1,2,...,K$ and $l\neq k$ is defined as:

$$\Delta\left(X_i, X_l\right)=\frac{\sum_{X_l \in C_l} d\left(X_i, X_l\right)}{\left|C_l\right|}. \tag{24}$$

The cluster with the lowest average dissimilarity is designated the *neighboring cluster* of X_i, as it is the best cluster for this object, with the exception of cluster C_k.

Let $b\left(X_i\right)$ be the smallest average distance $\Delta\left(X_i, X_l\right)$ computed for object X_i. This value indicates the distance between X_i and its neighboring cluster. Using Eq. (24), it is calculated as follows:

$$b\left(X_i\right)=\min_{\substack{l=1,2,..K \\ l \neq k}}\left\{\frac{1}{\left|C_l\right|} \sum_{X_l \in C_l} d\left(X_i, X_l\right)\right\}. \tag{25}$$

It can be said that while $a\left(X_i\right)$ represents the mean distance between X_i and other objects within the same cluster, $b\left(X_i\right)$ signifies the minimum of the mean distances between X_i and objects within other clusters. Then silhouette width (*sw*) of object X_i is calculated according to the following formula:

$$sw\left(X_i\right)=\frac{b\left(X_i\right)-a\left(X_i\right)}{\operatorname{Max}\left\{a\left(X_i\right), b\left(X_i\right)\right\}}, \tag{26}$$

where $-1 \leq sw\left(X_i\right) \leq 1$. In the event that cluster C_k comprises a single object, it is not evident how $a\left(X_i\right)$ should be defined. In such instances, we simply set $sw\left(X_i\right)$ equal to zero. This choice is of course arbitrary; however, it is proposed that a value of zero represents the most neutral position. The silhouette coefficient SC of the whole clustered data set is defined as:

$$SC=\begin{cases}\frac{1}{p} \sum_{X_i \in \mathbf{X}}^{K} s\left(X_i\right), & \text{if } p \neq 1 \\ 0, & \text{if } p=1\end{cases}, \tag{27}$$

where $SC \in[-1.1]$. A value of SC approaching 1 indicates that the data point is well-clustered. A value approaching zero indicates that the clusters are in fact overlapping. A value of approximately (-1) indicates that the data point has been incorrectly classified. In this context, it can be said that the silhouette coefficients are an effective means of assessing the degree of separation achieved by a given cluster in relation to other clusters. In the event of two clustering results being produced from the same data set using different clustering algorithms or different parameters, the clustering result with the higher value of SC is to be regarded as the optimal result.

2.2.3. *Gap Statistics*

The selection of the appropriate number of clusters in *k*-means is of paramount importance, given its considerable impact on the outcomes. The Gap statistics [25, 29] is a methodology employed to address this issue by comparing the extent of variation in the data across different cluster numbers (*k*). This is achieved by utilizing varying cluster numbers and assessing the total within-cluster variation against its expected value under a distribution devoid of discernible clustering. The optimal clusters are identified as the value at which the gap statistic is maximized, indicating that the clustering structure significantly surpasses randomness.

The Gap statistic has been shown to provide a more objective basis for determining the optimal number of clusters than subjective methods such as trial and error or guesswork. By statistically evaluating how well the clusters are described against random distributions, it is possible to make more informed decisions that will improve model performance.

In the calculation of a gap statistic, a number of random samples are generated and subsequently clustered over a range of *k*. The resulting inertia is then recorded, thus allowing for a number of inertiae for random cases. The original data set is also clustered over a range of *k*, resulting in a series of inertiae. The gap statistic, at *k* clusters, is then calculated as follows:

$$Gap(k) = \frac{1}{B}\sum_{i=1}^{B} \log\left(W_k^{(i)}\right) - \log\left(W_k\right), \tag{28}$$

where B is the number of reference datasets used and $W_k^{(i)}$ is the within-cluster sum of squared errors(WSS) for the reference data from the ith random sample (i=1,2,…,B) with k clusters, and W_k is the WSS from the original data with k clusters. We also calculate its standard deviation as

$$s_k = \sqrt{\left(1+\frac{1}{B}\right)\frac{1}{B}\sum_{i=1}^{B}\left(log\left(W_k^{(i)}\right)-\overline{W}\right)^2}, \tag{29}$$

where $\overline{W}=\frac{1}{B}\sum_{i=1}^{B}\log\left(W_k^{(i)}\right)$. Then the optimal number of clusters is chosen as the smallest value of k such that

$$Gap(k) \geq Gap(k+1) - s_{k+1}, \tag{30}$$

where s_{k+1} is the standard deviation of the Gap statistic for k+1 clusters. This condition ensures that the selected k provides a substantial enhancement over the random data configuration.

CHAPTER 3

SIMULATIONS AND RESULTS

This study analyses the micromorphology of Teucrium, encompassing a total of 40 species. All of the requisite analyses were conducted using the Mathematica 14.1 software. A comparison between these species was undertaken on the basis of their analyzed structural micromorphological characteristics, specifically those pertaining to their nuts and leaves. Table 1 illustrates these findings. The differentiation of these features in the Teucrium genus represents the most useful taxonomic characters. The identification of these features, whether absent or present, as well as their specific types, is crucial for taxonomic classification of this genus at the infra-generic level. In this context, a data matrix with 40 rows and 21 columns, which represents each of species and their different coded characters, has been constructed and is provided in Table 2 for reference.

Table 1: Descriptions of characters coded in a multistate format

Nutles	*ch1*	longitudinal ridges	*absent-0, present-1*
	ch2	surface ornamentation	*fovuloriate-0, striate,reticulate-1, reticulate-alveolate-2, reticulate-3, reticulate-foveate-4, alveolate-5, foveate-6*
	ch3	B type trichomes	*absent-0, present-1*
	ch4	A type trichomes	*absent-0, present-1*
	ch5	eglandular trichomes	*absent-0, present-1*
	ch6	thin-walled trichomes	*absent-0, present-1*
	ch7	F2 type trichomes	*absent-0, present-1*
	ch8	F4 type trichomes	*absent-0, present-1*
Leaves	*ch9*	B type trichomes	*absent-0, present-1*
	ch10	A1 type trichomes	*absent-0, present-1*
	ch11	A2 type trichomes	*absent-0, present-1*
	ch12	trichomes	*A1 type trichomes-0, A2 type trichomes-1, A1 and A2 types trichomes- 2*
	ch13	thin-walled trichomes	*absent-0, present-1*
	ch14	F2 type trichomes	*absent-0, present-1*
	ch15	F5 type trichomes	*absent-0, present-1*
	ch16	thick-walled trichomes	*absent-0, present-1*
	ch17	G2 type trichomes	*absent-0, present-1*
	ch18	G3 type trichomes	*absent-0, present-1*
	ch19	O type trichomes	*absent-0, present-1*
	ch20	leaves shape	*linear-ovate-0, linear-lanceolate-1, linear-2, ovate-lanceolate-3, ovate-elliptic-4, ovate-5, ovate- orbicular-6*
	ch21	Leaves margin	*not revolute-0, revolute-1, usually revolute-2*

Table 2: Data matrix with micromorphological characteristics of species of Teucrium

Species	*Abbreviations*	Characters																				
		ch1	*ch2*	*ch3*	*ch4*	*ch5*	*ch6*	*ch7*	*ch8*	*ch9*	*ch10*	*ch11*	*ch12*	*ch13*	*ch14*	*ch15*	*ch16*	*ch17*	*ch18*	*ch19*	*ch20*	*ch21*
T. creticum	cret	1	1	1	0	1	1	1	0	1	0	1	0	1	0	1	1	0	0	0	1	1
T. sandrasicum	sand	1	1	1	0	0	0	0	0	0	0	1	0	0	0	0	1	0	0	0	1	1
T. brevifolium	brev	1	1	1	0	1	1	0	1	0	0	1	0	0	0	0	1	1	0	0	1	1
T. pestalozzae	pest	1	1	1	0	1	1	0	1	1	0	1	0	0	0	0	1	1	0	0	1	1
T. ekimii	ekim	1	1	1	0	1	1	0	1	1	0	1	0	0	0	0	1	0	0	0	3	1
T. alyssifolium	alys	0	2	1	0	1	1	1	0	1	0	1	0	0	0	0	1	0	0	0	4	1
T. pseudaroanium	pseu	1	1	1	0	1	1	1	0	1	0	1	0	0	0	0	1	1	0	0	4	1
T. multicaule	multi	0	3	1	0	1	1	1	1	1	0	1	0	0	0	0	1	1	0	0	2	1
T. orientale var. orientale	orin	0	3	1	0	1	1	1	1	1	0	1	0	1	2	0	0	0	0	0	2	1
T. orientale* var. *puberulens	pube	0	3	1	0	1	1	0	1	1	0	1	0	1	0	0	0	0	0	0	2	1
T. orientale* var. *glabrescens	glab	0	4	1	1	1	1	1	0	1	0	1	0	1	0	0	0	0	0	0	2	1
T. pruinosum	prui	0	4	1	0	1	1	1	1	1	0	1	0	1	0	1	0	0	0	0	2	1
T. parviflorum	parv	0	4	1	1	1	1	1	1	1	0	1	0	1	1	0	0	0	0	0	1	1
T. scordium* ssp. *scordioides	scdi	0	5	1	0	0	0	0	0	1	1	0	1	1	0	1	0	0	0	0	4	0
T. melissoides	meli	0	5	1	0	1	1	1	0	1	1	1	0	1	1	1	0	0	0	0	4	0
***T. chamaedrys* subsp. *chamaedrys* L.**	cham	0	2	0	0	0	0	0	0	1	1	0	1	0	0	0	1	0	1	0	4	1

Species	Abbreviations	ch1	ch2	ch3	ch4	ch5	ch6	ch7	ch8	ch9	ch10	ch11	ch12	ch13	ch14	ch15	ch16	ch17	ch18	ch19	ch20	ch21
T. chamaedrys subsp. *lydium*	lydi	0	2	0	0	0	0	0	0	1	0	1	0	0	0	0	1	1	0	0	4	1
T. chamaedrys ssp. *trapezunticum*	trap	0	2	1	0	0	0	0	0	1	0	1	0	0	0	0	1	0	1	0	4	1
T. chamaedrys. ssp. *tauricolum*	taur	0	2	0	0	0	0	0	0	1	0	1	0	0	0	0	1	1	0	0	4	1
T. chamaedrys ssp. *syspirense*	sysp	0	2	0	0	0	0	0	0	1	0	1	0	0	0	0	1	1	1	0	4	1
T. chamaedrys ssp. *sinuatum*	sinu	0	2	1	0	0	0	0	0	1	0	1	0	0	0	0	1	1	1	0	4	1
T. divaricatum ssp. *divaricatum*	diva	0	2	1	0	0	0	0	0	1	0	1	0	0	0	0	1	1	0	0	4	1
T. divaricatum ssp. *graecum*	grae	0	2	1	0	0	0	0	0	1	0	1	0	0	0	0	1	1	0	0	4	1
T. flavum ssp. helenicum	flav	0	2	0	0	0	0	0	0	1	0	1	0	0	0	0	1	1	0	0	4	1
T. leucophyllum	leuc	0	2	1	0	0	0	0	0	1	1	0	1	0	0	0	1	1	1	0	3	1
T. microphyllum	micr	0	2	1	0	0	0	0	0	1	0	1	0	0	0	0	1	1	0	0	4	1
T. montanum ssp. montanum	mont	0	6	0	0	0	0	0	0	1	0	1	0	0	0	0	1	0	0	0	1	1
T. polium ssp. polium	poli	0	6	0	0	0	0	0	0	1	1	0	1	0	0	0	1	0	0	1	1	1
T. montbretii ssp. montbretii	mntb	0	3	1	0	1	0	0	0	1	1	1	0	0	0	0	1	1	0	0	5	2
T. montbretii ssp. pamphylicum	pamp	0	3	1	0	0	0	0	0	1	1	1	0	0	0	0	1	1	1	0	5	2
T. montbretii ssp. yıldırımlii	yild	0	3	1	0	0	0	0	0	1	1	1	1	0	0	0	1	0	1	0	5	2
T. odontites	odon	0	3	1	0	0	0	0	0	1	0	1	0	1	1	0	0	0	0	0	6	2
T. cavernarum	cave	0	3	1	0	0	0	0	0	1	1	0	1	1	1	1	0	0	0	0	5	2

T. antitauricum	anti	*0*	*3*	*1*	*0*	*0*	*0*	*0*	*0*	*1*	*1*	*0*	*1*	*1*	*1*	*0*	*0*	*0*	*0*	*0*	*5*	2
Species	*Abbreviations*	*ch1*	*ch2*	*ch3*	*ch4*	*ch5*	*ch6*	*ch7*	*ch8*	*ch9*	*ch10*	*ch11*	*ch12*	*ch13*	*ch14*	*ch15*	*ch16*	*ch17*	*ch18*	*ch19*	*ch20*	*ch21*
T. paederotoides	paed	*0*	*3*	*1*	*0*	*0*	*0*	*0*	*0*	*1*	*1*	*1*	*1*	*0*	*0*	*0*	*1*	*0*	*1*	*0*	*5*	*2*
T. lamiifolium. ssp. lamiifolium	lami	*0*	*4*	*1*	*0*	*1*	*1*	*0*	*1*	*1*	*1*	*0*	*1*	*0*	*0*	*0*	*1*	*0*	*1*	*0*	*5*	*0*
T. lamiifolium ssp. stachyophyllum	stac	*0*	*4*	*1*	*0*	*0*	*0*	*0*	*0*	*1*	*1*	*0*	*1*	*0*	*0*	*0*	*1*	*0*	*1*	*0*	*5*	*0*
T. hircanicum	hirc	*0*	*4*	*1*	*0*	*1*	*1*	*0*	*1*	*1*	*0*	*1*	*0*	*1*	*1*	*1*	*0*	*0*	*0*	*0*	*5*	*0*
T. aladaghensis	alad	*0*	*4*	*1*	*0*	*0*	*0*	*0*	*0*	*1*	*0*	*1*	*0*	*1*	*0*	*0*	*0*	*0*	*0*	*0*	*5*	*0*
T. kotschyanum	kots	*0*	*0*	*1*	*0*	*1*	*0*	*0*	*0*	*1*	*0*	*1*	*0*	*0*	*0*	*0*	*1*	*1*	*0*	*0*	*0*	*0*

Based on this data matrix, we try to classify the species of *Teucrium* by evaluating the micromorphological variability among them. Firstly, the distances between all other pairs of objects can be calculated and recorded in a matrix, which is known as the distance matrix. This matrix contains non-diagonal elements that represent the distances between pairs of objects, whereas the diagonal elements are set to zero. Secondly, we generate the correlation matrix for scaled *Teucrium* data to identify how strongly correlated two or more variables are. The computation of the correlation matrix allows for the discernment of the interrelationships among variables. These interrelationships may be understood as how the variables vary with regard to one another in terms of the mean. Based on the provided information, it can be concluded that the matrix represents a summary of the intercorrelations between all potential combinations of variables. If we want to explore correlations in Teucrium data, we can visualize the correlation matrix by using correlogram in Fig.1. A green display shows there is a positive correlation, while a red display indicates there is a negative one. The degree of color intensity and the diameter of the circle are directly correlated with the corresponding correlation coefficient.

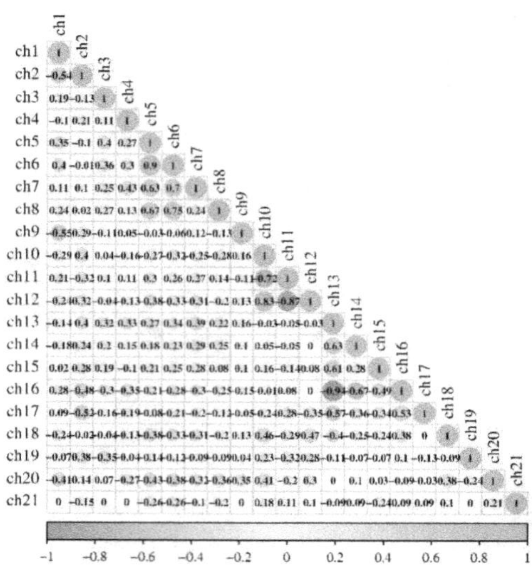

Figure 1: Visualize the correlation matrix of Teucrium data using corrplot

By utilizing the correlation values between the variables illustrated in Figure 1, we obtain a KMO value of 0.619. The observed value exceeds the recommended minimum of 0.5, indicating that a particular data set from Teucrium would be suitable for factor analysis.

The proportion of the total variation in the original data that is accounted for by the ith factor and the first m number of factors is defined as $\frac{\lambda_i}{\sum_{j=1}^{p}\lambda_j}\times 100$ and $\frac{\lambda_1+\lambda_2+\ldots+\lambda_m}{\sum_{k=1}^{p}\lambda_k}\times 100$. Table 3 clearly shows that the total variance achieved for two factors is 45%. In addition, their corresponding eigenvalues 5.242 and 4.196 are greater than others. This can be helpful in selecting the final factors for dimension reduction.

Table 3: Ordered eigenvalues of correlation matrix and corresponding percentages of total variability for Teucrium data

Variables	Eigenvalues	Percent of Variances	Cumulative Percent
ch1	5.242	25	25
ch2	4.196	20	45
ch3	1.928	9	54
ch4	1.724	8	62
ch5	1.366	7	69
ch6	1.206	6	75
ch7	0.903	4	79
ch8	0.874	4	83
ch9	0.634	3	86
ch10	0.568	3	89
ch11	0.466	2	91
ch12	0.463	2	93
ch13	0.382	2	95
ch14	0.293	2	97
ch15	0.221	1	98
ch16	0.186	1	99
ch17	0.160	1	100
ch18	0.079	0	100
ch19	0.045	0	100
ch20	0.028	0	100
ch21	0.025	0	100

In factor analysis, the intercorrelations between the variables and factors are expressed as factor loadings, which are also known as component loadings. The first two factor loadings and communalities for variable $\{ch_i\}_{i=1}^{21}$ are given in Table 4, while a bar chart illustrating them is presented in Fig.2.

Table 4: First two factor loadings and graphically rotated loadings for sake of simplicity and communalities for standardized variables

Variables	Unrotated Loadings l_{i1}	l_{i2}	Rotated Loadings l_{i1}^*	l_{i2}^*	Communality $l_{i1}^2+l_{i2}^2$	Percent explained
ch1	0.061	-0.128	0.007	-0.141	0.020	2.015
ch2	0.001	0.178	0. 069	0. 164	0.031	3.170
ch3	0.078	0.026	0. 082	−0.005	0.006	0.689
ch4	0.078	0.037	0.086	0.004	0.007	0.757
ch5	0.156	-0.028	0.133	−0.086	0.025	2.534
ch6	0.163	-0.009	0.147	−0.071	0.026	2.680
ch7	0.135	0.022	0.134	−0.031	0.018	1.898
ch8	0.122	-0.011	0.108	−0.057	0.015	1.507
ch9	-0.020	0.093	0.016	0.094	0.009	0.921
ch10	-0.099	0.142	−0.037	0.170	0.030	3.033
ch11	0.083	-0.139	0.024	−0.161	0.026	2.652
ch12	-0.104	0.140	−0.043	0.169	0.030	3.070
ch13	0.114	0.158	0.166	0.102	0.038	3.840
ch14	0.078	0.124	0.120	0.085	0.021	2.173
ch15	0.067	0.117	0.107	0.082	0.018	1.830
ch16	-0.102	-0.167	−0.159	−0.115	0.038	3.880
ch17	-0.045	-0.160	−0.103	−0.131	0.027	2.795
ch18	-0.114	0.017	−0.099	0.059	0.013	1.339
ch19	-0.041	0.042	−0.021	0.055	0.003	0.351
ch20	-0.088	0.085	−0.049	0.113	0.015	1.519
ch21	-0.047	-0.010	−0.048	0.008	0.002	0.239

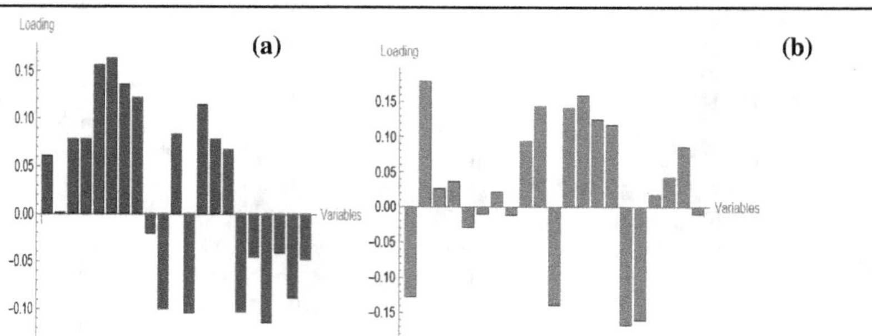

Figure 2. Barchart of first two factor loadings: (a) Loading 1 (b) Loading 2

The coefficients of loading 1 (l_{i1}) and loading 2 (l_{i2}) are given in columns 2 and 3 of Table 4.The loadings of the variable *ch1* on F_1 and F_2 are $l_{11} = 0.061$ and $l_{12} = -0.128$, respectively. The higher the loading, the more strongly the observed variable is related to the factor. Table 4 and Fig. 1 demonstrate that *ch2* has a negligible impact on the explanation of the variation observed in F_1 whereas *ch6* is highly correlated with $F_1$and F_2. On the other hand, *ch2* variable has the highest loading (0.178) on F_2. Therefore, *ch6* and *ch2* are of pivotal importance in elucidating the observed variation on F_1 and F_2, respectively. As illustrated in Table 4, the variance in *ch13* and *ch16* is accounted for by 0.38%.

The objective of factor analysis is to discern and extract "*easily interpretable common factors*" from the data set. However, the initial analysis results are difficult to interpret. The factor ratio is a process through which more readily interpretable factors, referred to as rotated factors, can be obtained. This will permit the attribution of an interpretation to each factor that is derived from the variables with which it is most strongly correlated. It is not unusual to experience challenges in obtaining information from the original factor loadings. In such instances, factor rotation may be a necessary step. The term "factor rotation" understood to signify the rotation of the vertical axes, either in a clockwise or anticlockwise direction, by a specified angle. The aim of this procedure is twofold: firstly, to guarantee that the factor loadings are uncorrelated, and secondly, to optimize the value of the factor loadings. In order to attain this objective, the factor loadings must be as close as is feasible to one of the vertical axes, while ensuring that the data are orthogonalised. In this instance, the rows of **L** represent pairs of loadings (l_{i1}, l_{i2}), $i = 1, 2, ..., p$, corresponding to $x_1, x_2, ..., x_p$. An angle, designated as θ, is selected for rotation of the axes in order to facilitate closer alignment with the groupings of points. The novel rotated loadings (l_{i1}^*, l_{i2}^*)

can be expressed as coordinates on the graph directly or alternatively, they can be calculated by applying the formula $\mathbf{L}^* = \mathbf{L}.\mathbf{T}$ using

$$\mathbf{T} = \begin{pmatrix} \cos\theta & -\sin\theta \\ \sin\theta & \cos\theta \end{pmatrix}, \tag{31}$$

where $\mathbf{T}$ is a orthogonal matrix $\mathbf{T}^T\mathbf{T} = \mathbf{T}\mathbf{T}^T = \mathbf{I}$. By employing the values of $\mathbf{L}$ from Table 4 and 22.5° in $\mathbf{T}$ as indicated in Eq. (31), we can calculate the rotated loadings presented in Table 4. For purposes of comparison, the original unrotated loadings and the graphically rotated loadings for Teucrium data are set forth in Table 4. The 21 pairs of loadings (l_{i1}, l_{i2}) and rotated loadings (l_{i1}^*, l_{i2}^*) corresponding to the 21 variables are clearly displayed in Fig. 3.

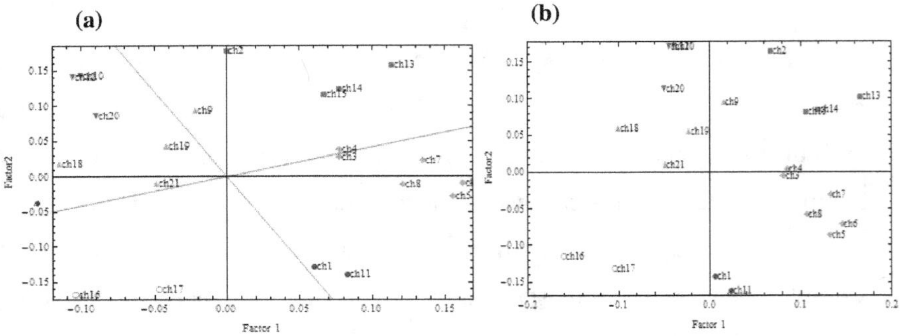

Figure 3: Rotation of loadings illustrated (a) Rotating the coordinate axes anticlockwise by 22.5° (b) The new coordinate axes indicated by the red lines

An orthogonal rotation through 22.5° would result in the axes (factors) being positioned in closer alignment with the two clusters of points (variables). The rotation would serve to enhance the degree of correspondence between each cluster of variables and a factor.

In order to perfectly reproduce an observed correlation matrix among k measured variables, k factors are required. However, this doesn't simplify the dimensionality. Thus, we need to ask how many factors are necessary for reproducing the observed correlation matrix reasonably well? This is usually done by different criteria, such as Kaiser's and Cattell's Criteria [30]. Cattell proposed a graphical approach as the most straightforward method, which is represented by a scree plot. This is a graph of the eigenvalues in relation to the component numbers. The eigenvalue serves to quantify the extent to which a factor accounts for the observed variance. Basically, the ordered eigenvalues versus component number are plotted sequentially and shown in Fig. 4.

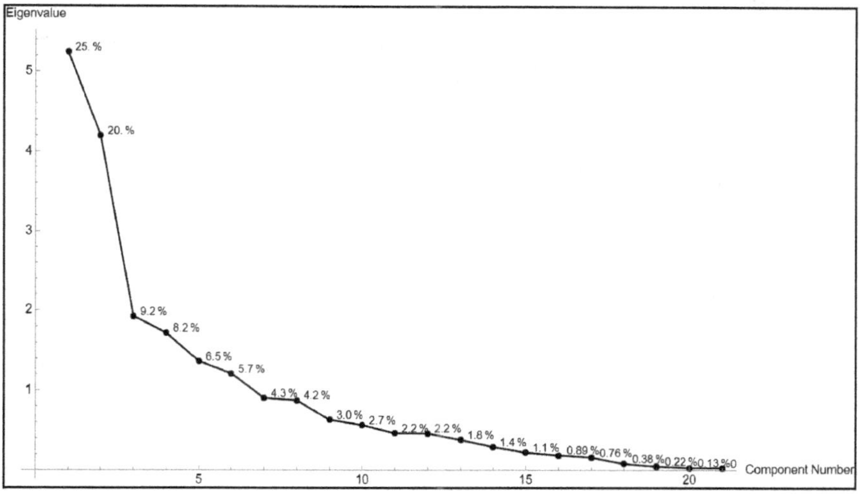

Figure 4: Scree plot for the correlation matrix for data that most probable have two fundamental factors

As demonstrated in Fig. 4, the initial two eigenvalues exhibit a pronounced curvature, followed by a bend and then a linear trend with a relatively shallow slope. Moreover, a notable decline can be observed between the second and third factors. Thus, we decide the number of factor scores or optimal number of components m should be 2. Once the number of factors to be included has been determined, the factor scores are calculated. This is achieved by taking the standardized score for each variable, multiplying it by the corresponding factor loading for the specified factor, and then summing these products.

The eigenvalue attributed to a particular component is equivalent to the sum of the squares of the factor scores associated with that component. Consequently, the significance of an observation in relation to a specific component can be quantified using the following ratio:

$$ctr_{i,l} = \frac{f_{i,l}^2}{\lambda_l}, \quad (32)$$

where $f_{i,l}$ is the factor score of observation i to component l and λ_l is the eigenvalue of the lth component. The ratio, which ranges between 0 and 1, may also be designated as the "contribution of the observation to the component". It can be confidently asserted that the total of all the observed contributions is equal to 1. The significance of an observation's contribution to a component can be quantified by measuring its magnitude, which is directly proportional to the extent of that contribution.

Fig. 4 states that we can now plot each Teucrium species' score across all subjects in a 2D graph. Accordingly, the factor scores for the initial two components, along with the observed contributions to said components, are presented in Table 5. The significant contributions are indicated by bold text. For reasons of convenience, the contributions have been multiplied by a hundred and subsequently approximated to the nearest integer.

Table 5: Factor scores and observed contributions to the components

Species	FScore 1	FScore 2	ctr1	ctr2
cret	1.355	-0.420	5	0
sand	0.052	-1.478	0	6
brev	0.885	-1.906	2	**9**
pest	0.791	-1.482	2	6
ekim	0.772	-1.050	2	3
alys	0.590	-0.436	1	0
pseu	0.667	-1.242	1	4
multi	0.889	-0.765	2	2
orin	1.784	0.792	**8**	2
pube	1.123	0.202	3	0
glab	1.523	0.577	6	1
prui	1.633	0.706	7	1
parv	2.027	0.766	**11**	**2**
scdi	-0.299	1.975	0	**10**
meli	1.283	1.554	4	6
cham	-1.498	0.524	6	1
lydi	-0.672	-0.804	1	2
trap	-0.637	-0.371	1	0
taur	-0.672	--0.804	1	2
sysp	-0.933	-0.765	2	2
sinu	-0.729	--0.695	1	1

Table 5. (Continued)

Species	FScore 1	FScore 2	ctr1	ctr2
diva	-0.468	-0.735	1	1
grae	-0.468	-0.735	1	1
flav	-0.672	-0.804	1	2
leuc	-1.330	0.216	5	0
micr	-0.468	-0.735	1	1
mont	-0.411	-0.129	0	0
poli	-1.327	1.106	5	3
mntb	-0.500	-0.327	1	0
pamp	−1.074	-0.230	3	0
yild	-1.221	0.414	4	0
odon	0.066	0.772	0	2
cave	-0.346	2.008	0	**10**
anti	-0.534	1.684	1	7
paed	-1.221	0.414	4	0
lami	-0.343	0.817	0	2
stac	-1.268	0.920	4	2
hirc	1.394	1.101	5	3
alad	0.112	0.612	0	1
kots	0.144	-1.246	0	4

By utilizing the factor score values outlined in Table 5, it is possible to construct scatter plots of Teucrium species in two dimensions. It is a requirement of some clustering algorithms, including K-means [31], that the number of clusters is predetermined prior to the classification process. To find the best number of clusters for the Teucrium data set, the average silhouette scores are calculated for each cluster number. Figure 5 shows the silhouette diagrams obtained by changing the cluster number, *k*, from three to seven. Figure 5 (d) illustrates that, for a predefined number of clusters, *k*=6 produces cluster of a similar size to those produced by other values of *k*. This is evidenced by the superior silhouette values observed for *k*=6, which indicates that the points within these clusters are more effectively separated than in other cases. In light of the preceding analysis, the optimal number of clusters is determined to be *k*=6.

A variety of methods and indexes are employed in the literature to establish the optimal number of clusters within a given data set, including average Silhouette(S), Gap statistics , Elbow [31], Dendrogram [32], RSquared (RS) [33], Caliński-Harabasz (CH) [34], Davies-Bouldin (DB) [35] and Dunn Index (D) [36].

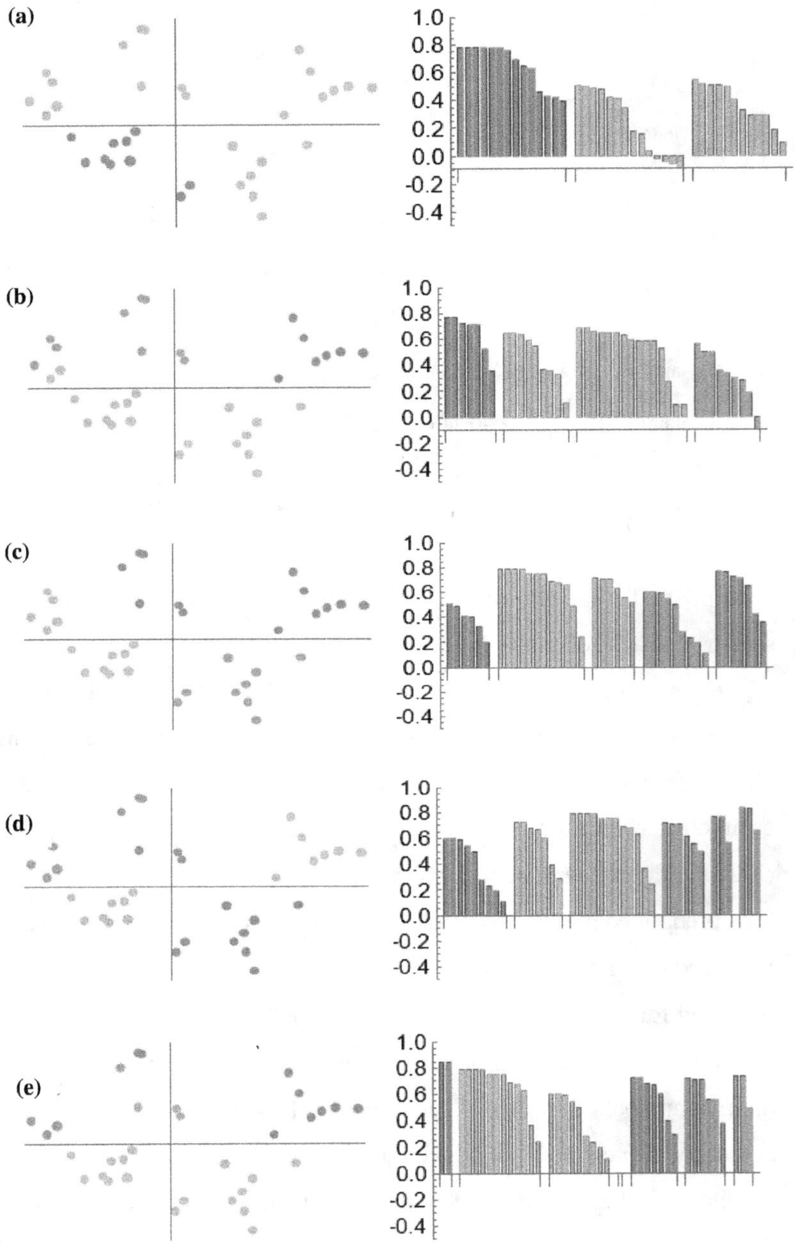

Figure 5:Silhouette analysis of clustering on Teucrium species data with varying k clusters from 3 to 7. (a) k=3 clusters (b) k=4 clusters (c) k=5 clusters (d) k=6 clusters (e) k=7 clusters

A comparison of the silhouette method with the most widely employed methods in the existing literature to find the best number of clusters is shown in Figure 6. These include the gap method, the elbow method and the dendrogram method, which is frequently employed in hierarchical classification for the classification of Teucrium species data. It is evident that the silhouette coefficient attains its maximum value when there are six clusters. The silhouette plots suggest that a division into k = 6 clusters is probably most natural. The Gap statistical method is also compatible with the silhouette method and indicates that Teucrium species can be grouped into six clusters. In contrast to the silhouette method, the elbow and dendrogram methods indicated that the Teucrium species can be classified into five and seven distinct clusters, respectively (see Figure 6). The Elbow method, which is widely utilized due to its simplicity, has the potential to yield misleading detections in complex data distributions [37]. It requires running the clustering algorithm with different values of k and calculating the within-cluster sum of squared errors (WCSS) for each number of k. The WCSS is calculated according to the following mathematical formula:

$$WCSS = \sum_{i=1}^{k} \sum_{j=1}^{n_i} \left\| x_j - \mu_i \right\|^2 \tag{33}$$

where x_j represents a data point within cluster i, while the variable μ_i denotes the centroid of cluster i. The variable n_i signifies the number of data points within cluster i, and k represents the total number of clusters. Once the WCSS values have been calculated in accordance with Eq. (33), a curve is plotted in alignment with the specified number of clusters, designated as k. The position of a knee in the plot is commonly accepted to be an indicator of the k_{opt}. This value is determined via a statistical analysis process.

Upon plotting WCSS against k for the Teucrium dataset, the resulting graph exhibits a resemblance to an elbow, as illustrated in Fig. 6. As the number of clusters is increased, the WCSS value begins to decrease. As expected, the plot resembles an arm with a clear elbow at k = 5. Therefore, we can infer that the optimal number of clusters is $k_{opt} = 5$. If we have a large dataset, it can become difficult ascertain the most appropriate number of clusters by the dendrogram. Therefore, the number of clusters could not be ascertained automatically. Instead, it had to be determined through an intuitive process and the elbow method was used. Conversely, the 'elbow method' is a heuristic approach to determining the optimal number of clusters. Consequently, this method may not yield an objective result.

The application of the dendrogram method [38], a form of hierarchical agglomerative clustering, can present certain difficulties, contingent on the nature of the data being analyzed. Typically, the data must be pre-processed. One of the intrinsic limitations of dendrogram-based techniques is the lack of information pertaining to the number of clusters and the most appropriate threshold for dendrogram construction, which are essential for the creation of clusters. Thus, the number of clusters could not be ascertained automatically. Instead, it had to be determined through an intuitive process [39].

The gap statistic is a widely used method among researchers to determine the optimal number of clusters. Nevertheless, this approach may occasionally produce inconsistent results due to the inherent challenges involved. In Figure 6, the Gap statistical method was executed for 500 iterations. Should the number of iterations be reduced, for example to 100, the Gap statistics method may yield different values for the optimal number of clusters. These include sensitivity to outliers, the overlap between data points, and the choice of the null reference distribution. These issues have the potential to significantly impact the efficacy of the Gap statistic, particularly in datasets with heterogeneous attributes and complex configurations [40].

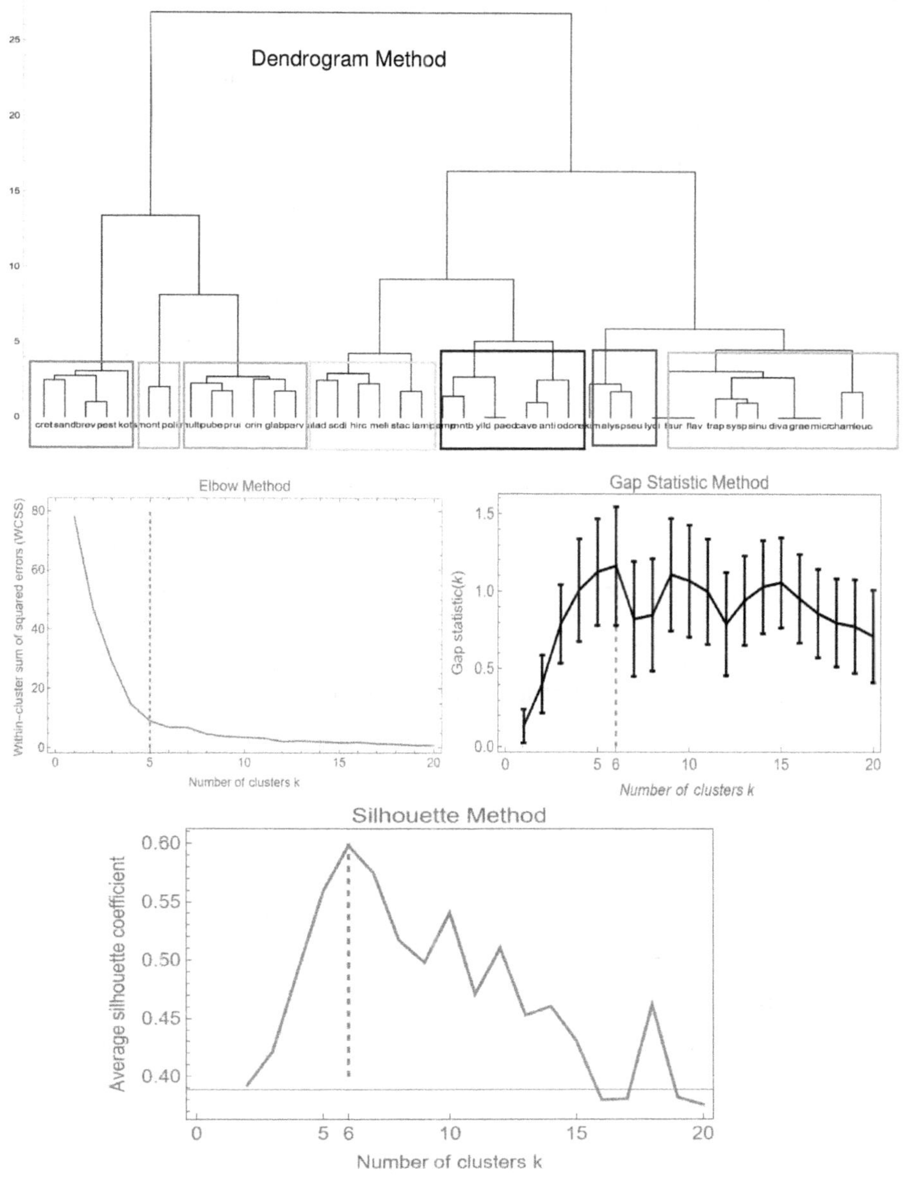

Figure 6: A graphical comparison of the most commonly employed optimal number of clusters estimators for classifying Teucrium data

A comparative analysis of the performance of the average Silhouette, Davies-Bouldin, Caliński-Harabasz and Dunn Index methods [41,42] is presented in reference [43]. The findings indicate that the average Silhouette method yielded superior results. Table 6 illustrates the diverse optimal cluster number estimator index values of clustering methods, including Silhouette, Calinski-Harabasz, Davies-Bouldin, Dunn and R-Squared, for a range of clustering methods, such as K-Means, K-Medoids [44], Agglomerate [45], DBSCAN [46], Jarvis-Patrick [47], Spanning Tree [48], Spectral [49], Density peak cluster [50] and Feature weighting K-Means [51] which have been employed for the classification of Teucrium species.

Table 6: The optimal cluster number estimator index values for some clustering methods when used with Teucrium species data

Clustering Method		Optimal number of clusters estimator index and method				
	k_{opt}	S	CH	DB	D	RS
K-Means	6	0.598	69.898	0.497	0.306	0.732
Density Peak Cluster	7	0.562	73.220	0.538	0.343	0.753
K-Medoids	5	0.558	66.326	0.614	0.306	0.683
Feature Weighting K-Means	10	0.502	69.889	0.652	0.160	0.810
Agglomerate	16	0.422	91.472	0.605	0.323	0.884
DBSCAN	16	0.400	80.022	0.365	0.501	0.883
Jarvis-Patrick	2	0.400	25.095	1.085	0.149	0.247
Spanning Tree	7	0.369	28.212	0.614	0.269	0.638
Spectral	12	0.392	37.946	0.933	0.061	0.815

The average Silhouette index assumes values between -1 and 1. A higher value of this index signifies that the dataset is clustered with greater efficacy. As illustrated in Table 6, the K-Means method has the highest silhouette value, with a value of approximately 0.6. The other two clustering methods with the highest S values are Density peak cluster and K-Medoids methods. The Davies-Bouldin index is defined such that its values are restricted to the range 0 to 1. A lower value of this index is an indication that the dataset is clustered in an optimal manner. As demonstrated in Table 6, the lowest values of this index are observed in the DBSCAN, K-Means and Density peak cluster methods. Furthermore, it has been observed that the DB index of the K-Means method is lower than that of the

Density peak cluster method. The higher the values of the Caliński-Harabasz and Dunn indices, the more effective the clustering of the dataset. Moreover, the DBSCAN, Agglomerate, K-Means and Density peak cluster methods exhibit the highest values for these indices. The CH values of the K-Means and Feature weighting K-Means methods are approximately 70 and are found to be almost identical. However, the Dunn index (D) value of the Feature weighting K-Means method is 0.16, which is considerably lower than the D value of 0.306 of the K-Means method. On the other hand, the CH value of the Density peak clustering method is higher than the values of the K-Means, K-Medoids, and Feature-weighted K-Means methods.

Table 7 illustrates the values of five different clustering validation measures, contingent upon the number of clusters, with four distinct clustering methods applied to the Teucrium species data set. It is evident that the RS index values demonstrate a monotonic increase as the number of clusters is augmented for all cluster methods. On the other hand, the remaining four indices attain their maximum or minimum value as the number of clusters is equivalent to the optimal number of clusters that can effectively classify the Teucrium data. For example, the silhouette index(S) and Davies-Bouldin (DB) attain their maximum value (0.598 and 0.306) when k is 6, as identified through the K-means clustering method. This indicates that the Teucrium species can be effectively classified into six optimal clusters according to their morphological characteristics. The alternative classification method, designated as the Agglomerate classification method, indicates that the CH, DB and D indices suggest that the Teucrium species can be classified into nine distinct clusters. In accordance with the K-Medoids clustering method, each classification index exhibits a distinct preference for the number of clusters. The implementation of the Spanning Tree method yields a prediction of nine clusters for the S and D indices, whereas the CH and DB indices suggest a total of ten clusters. In the context of the classification of Teucrium species based on their micromorphological characteristics, the Density peak cluster method proposes the identification of seven clusters for the S and CH indices. Conversely, the Feature weighting K-Means clustering method suggests ten clusters for the same indices.

Table 7: Experimental results of five clustering validation measures on Teucrium species data

Cluster Method	Clustering validation measures	Number of cluster(k) 2	3	4	5	6	7	8	9	10
KMeans	S	0.392	0.421	0.490	0.558	**0.598**	0.574	0.517	0.497	0.540
	CH	24.98	31.57	51.38	66.32	69.89	57.37	73.03	76.07	70.49
	DB	1.118	0.801	0.712	0.614	0.497	**0.489**	0.633	0.650	0.541
	D	0.134	0.147	0.158	0.306	**0.306**	0.222	0.23	0.284	0.284
	RS	0.256	0.446	0.586	0.683	0.732	0.741	0.766	0.795	0.813
K-Medoids	S	0.401	0.478	0.503	**0.558**	0.528	0.555	0.53	0.497	0.514
	CH	24.79	42.53	53.63	66.32	65.83	66.49	72.55	76.07	68.94
	DB	1.04	0.758	0.667	0.614	0.634	0.542	0.543	0.65	**0.508**
	D	0.044	0.199	0.12	0.306	0.209	0.307	**0.395**	0.284	0.205
	RS	0.262	0.463	0.601	0.683	0.708	0.746	0.769	0.795	0.804
Feature Weighting K-Means	S	0.377	0.389	0.461	0.410	0.478	0.442	0.401	0.437	**0.502**
	CH	23.49	30.98	40.240	31.017	54.17	50.379	45.15	48.95	**69.98**
	DB	1.171	0.831	0.653	0.803	0.801	0.840	0.823	0.72	0.652
	D	0.145	0.151	0.208	0.187	**0.307**	0.160	0.160	0.148	0.160
	RS	0.240	0.406	0.543	0.554	0.688	0.710	0.730	0.760	0.810
Density Peak Cluster	S	0.322	0.269	0.428	0.461	0.398	**0.562**	0.529	0.507	0.476
	CH	16.93	10.98	28.93	42.45	37.53	73.22	71.11	69.90	63.76
	DB	1.034	0.791	0.642	0.712	0.704	0.538	0.538	0.521	0.473
	D	0.147	0.147	0.172	0.194	0.194	0.343	0.343	0.311	0.311
	RS	0.180	0.262	0.514	0.641	0.663	0.753	0.773	0.791	0.804
Agglomerate	S	0.375	0.088	0.449	0.436	0.322	0.343	0.351	0.514	**0.52**
	CH	23.59	6.325	35.37	33.58	29.88	34.01	32.34	**74.29**	72.40
	DB	1.173	1.306	0.668	0.656	0.896	0.809	0.803	0.478	0.514
	D	0.063	0.129	0.28	0.18	0.106	0.119	0.119	0.284	0.226
	RS	0.243	0.182	0.522	0.593	0.591	0.652	0.682	0.791	0.811
SpanningTree	S	0.202	0.074	0.327	0.477	0.388	0.369	0.308	**0.484**	0.476
	CH	7.254	4.288	17.29	38.60	32.89	28.21	25.17	63.44	63.76
	DB	1.211	1.463	0.879	0.629	0.612	0.614	0.639	0.495	0.473
	D	0.242	0.237	0.229	0.327	0.285	0.269	0.257	**0.343**	0.311
	RS	0.109	0.159	0.419	0.605	0.624	0.638	0.658	0.785	0.804

The shaded areas in Table 7 show the clustering validation measures of each clustering method at the optimal number of clusters given in Table 6.

The performance of the classification methods (classifiers) is contingent upon six distinct complexity scenarios, contingent upon dimensionality and sample size [52]. This book provides a comparison of the graphical representation of the methods used for the classification of Teucrium species, rather than an analysis of the complexity scenarios themselves. Figure 7 presents a comparative graphical representation of several commonly employed classification techniques. The figure demonstrates that Teucrium species can be classified into six clusters using the K-Means method, five clusters using the K-medoids method, and seven clusters using the Density peak clustering and the Spanning Tree method. It is essential to highlight that the K-means, K-medoids, Density peak cluster and Feature weighting K-Means methodologies can only be employed when the number of clusters has been previously established. Conversely, DBSCAN, Agglomerate, Jarvis-Patrick, Spanning Tree and Spectral can only be used if the number of clusters is determined automatically. As illustrated in Figure 7, the K-means clustering method demonstrates a notable degree of separation and a relatively low level of dispersion in the classification of Teucrium species. It can thus be inferred that this constitutes an efficacious cluster for the data pertaining to the Teucrium species. Furthermore, the Jarvis-Petric and K-Medoids classifiers indicate a high degree of dispersion, whereas the Spectral, Spanning Tree and Agglomerate classifiers indicate a low level of separation. The application of the Density peak clustering method for the classification of Teucrium species demonstrates a relatively high degree of separation, which is similar to the K-Means method. However, the utilisation of the Feature weighting K-Means method results in a comparatively low level of separation and a low level of dispersion. On the other hand, Figure 7 shows that the DBSCAN classifier has a low level of separation and a high level of dispersion. Therefore, it can be concluded that this method is not effective in effectively clustering Teucrium species.

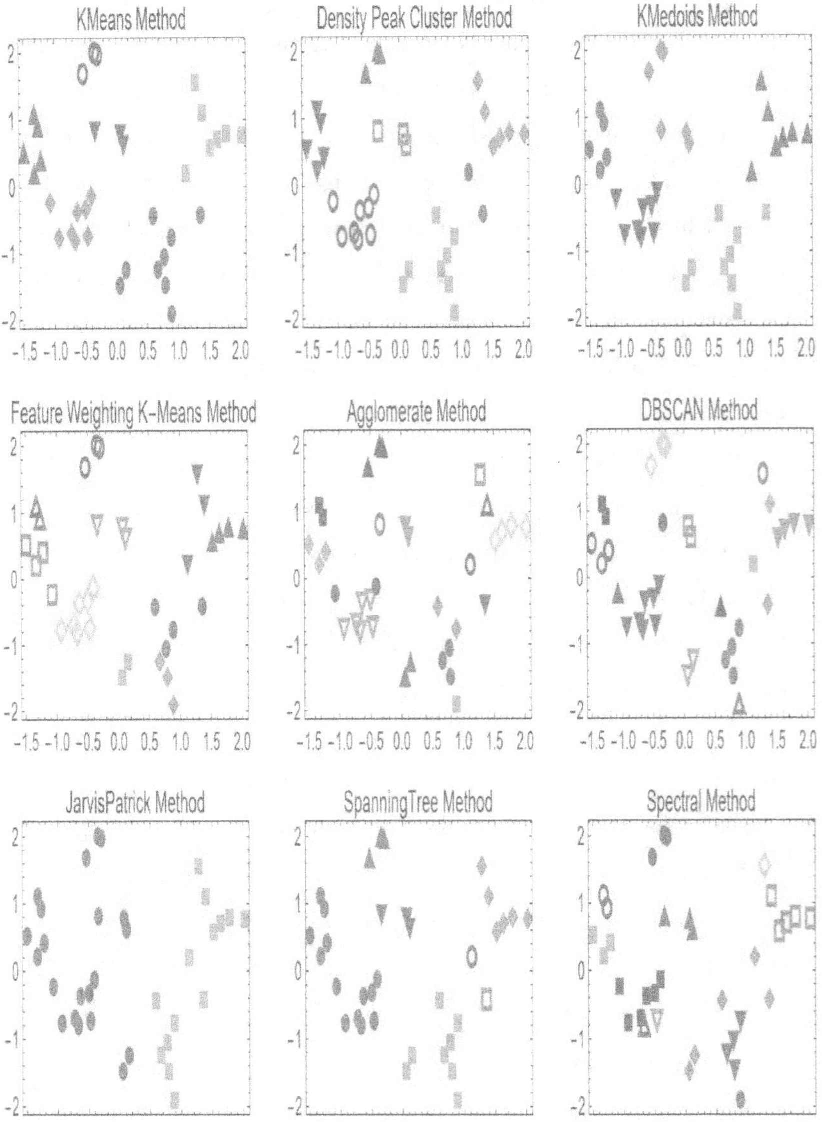

Figure 7: Classification results of common methods on the Teucrium species dataset

Figure 8 illustrates the two-dimensional and three-dimensional graphs of the classification of Teucrium species according to their micromorphological characters. The graphs have been created utilising the K-Means clustering method and the Silhouette optimal cluster number estimator index, which has been proposed in the present book. It has been demonstrated that the Teucrium species can be classified into approximately six functional categories.

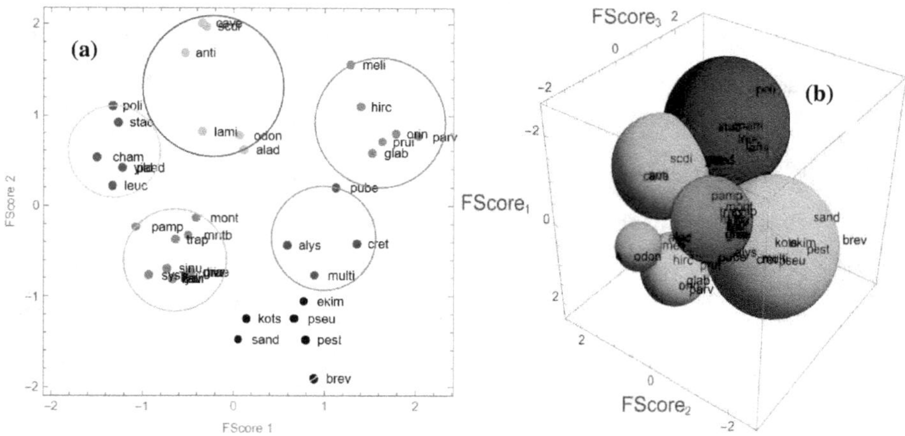

Figure 8: Plot the classification of Teucrium species in (a) two dimensions(2D) and (b) three dimensions (3D). The symbols in the figure legend are representative of the six functional classes. The full nomenclature of the abbreviated codes for Teucrium taxa is provided in Table 2.

CHAPTER 4

DISCUSSIONS AND CONCLUSION

In the present book we investigate how to classify Teucrium species according to their micromorphological characteristics encompassing the structural characteristics of leaves, nutlets, and trichomes. The data matrix comprises 40 species of Teucrium and 21 variables (cf. Tables 1 and 2). In order both to reduce the number of new variables in the dataset and to exploit the relationships between the original variables in order to reveal new structures (factors) that are not correlated but are assumed to be explained by the variables forming the group, factor analysis was employed. These factors provided a setting for the cluster analysis of Teucrium species. Therefore, this book presents an efficient algorithm [53] for the classification of multivariate data, which has been developed by combining factor analysis with cluster analysis and coded in the Mathematica programming language. This algorithm utilizes a silhouette coefficient to represent each cluster, which serves as a metric for cluster assessment. This coefficient is derived from a comparative analysis of the degree of cohesion and isolation exhibited by the cluster in question.

A comprehensive comparative analysis of our proposed methodology with several existing and commonly utilized techniques and approaches is presented. To determine the optimal number of clusters for the classification of Teucrium species, we present a comparison between the silhouette method and the dendrogram, the elbow and gap statistical methods that are commonly employed in existing literature. The average silhouette coefficients method, which is the proposed method, yielded an estimated number of clusters of six, while the gap statistical method supported the proposed method and identified this number of clusters as six (cf.Fig.6). In contrast, the elbow method and dendrogram methods, unlike the silhouette coefficient and gap statistical methods, yielded optimal cluster numbers of five and seven, respectively (cf.Fig.6). Although the interpretation of elbow and dendrogram plots is somewhat subjective, both the silhouette coefficient and gap statistics methods can determine the number of clusters with a great deal of accuracy. Similarly, the silhouette measure was compared with other clustering validation measures that are commonly used in this field, including Davies-Bouldin, RSquared, Caliński-Harabasz and Dunn Index (cf. Tables 6 and 7). The results of the empirical analysis and findings indicate that the average silhouette method is a reliable estimator of the number of true clusters in the classification process.

Computational experiments on Teucrium species data were utilized to assess the efficacy of various clustering methods and classification indices. Experimental results showed that the Jarvis-Patrick, K-Medoids, K-Means, Spanning Trees, Density Peak Cluster, Feature weighting K-Means, Spectral and DBSCAN clustering methods were able to classify the Teucrium species into 2, 5, 6, 7, 7, 10, 12 and 16 optimal clusters respectively, based on their micromorphological characteristics (cf. Table 6 and Fig.7). It is note that DBSCAN generally performs poorly on high-dimensional datasets. This is because, in such spaces, the distance between points tends to be more uniform and crowded. This can have an adverse effect on the performance of distance-based clustering algorithms such as DBSCAN. In high-dimensional spaces, the concept of density also becomes less meaningful because the points are more uniformly distributed. This causes DBSCAN's density-based approach to produce inaccurate cluster identification results. Dimensionality reduction is a common strategy for using DBSCAN effectively on high-dimensional data. In this book, we used the Factor analysis for dimension reduction. Nevertheless, it should be noted that the efficacy of this method is contingent upon two parameters, namely the minimum number of points (*minpts*) and the maximum distance (*eps*). Consequently, concerns may arise during the process of selecting the parameters of DBSCAN. As Table 6 and Figure 7 show, the performance of DBSCAN in clustering Teucrium species is lower than the performance of other clustering algorithms.

In the simulation studies, the feature weights were taken uniformly in the algorithm of the feature weighting K-Means method. However, it is important to note that the selection of feature weights may have a significant impact on the efficacy of the feature weighting K-Means method in terms of clustering performance. This issue will be addressed in future studies.

The Density peak clustering (DPC) algorithm has been a prevalent clustering method since 2014 [50]. This algorithm, defines clusters as regions around local maxima of density. Cluster centers, also referred to as density peaks in DPC, can be manually selected from a two-dimensional decision graph. The construction of a decision graph necessitates the utilization of two parameters, designated as $\boldsymbol{\rho}$ and $\boldsymbol{\delta}$. The parameter $\boldsymbol{\rho}$ is the local density, and $\boldsymbol{\delta}$ is the minimum distance between a given data point and another point with higher density [50,54]. As illustrated in Fig.9, the DPC algorithm is applied to the Teucrium species dataset for varying values of *dc*. In this context, "*dc*" is defined as the cutoff distance, a value that is frequently determined by extracting it from the Euclidean distance matrix of all data points based on a specified percentage. DPC utilizes the concepts of "$\boldsymbol{\rho}$" and "$\boldsymbol{\delta}$" to generate a decision graph, as illustrated in Fig. 9(a). The selection of cluster centers is predicated on the $\boldsymbol{\gamma} = \boldsymbol{\rho} \times \boldsymbol{\delta}$ top points. In the decision graph, points with high density and large distance are selected as cluster centers and illustrated in Fig. 9(b). The allocation of points is based on the proximity of each point to the nearest high-density cluster center. The clustering results obtained using the cluster centers in Fig. 9(b) are shown in Fig. 9(c). As illustrated in Fig. 9(c), the selection of cutoff distance values impacts the separation and dispersion levels in the clustering of Teucrium species. Consequently, the selection of *dc* values is a critical aspect of the DPC clustering algorithm.

In scenarios where the dimension of the dataset is relatively high and contains multiple density peaks within a cluster, the Density peak cluster algorithm faces difficulties in efficiently partitioning the cluster into multiple subsets. This results in a heightened time consumption during the clustering process. Consequently, DPC has poor clustering performance on high-dimensional datasets. For a detailed analysis of the merits and drawbacks of DPC's efficacy, refer to [54].

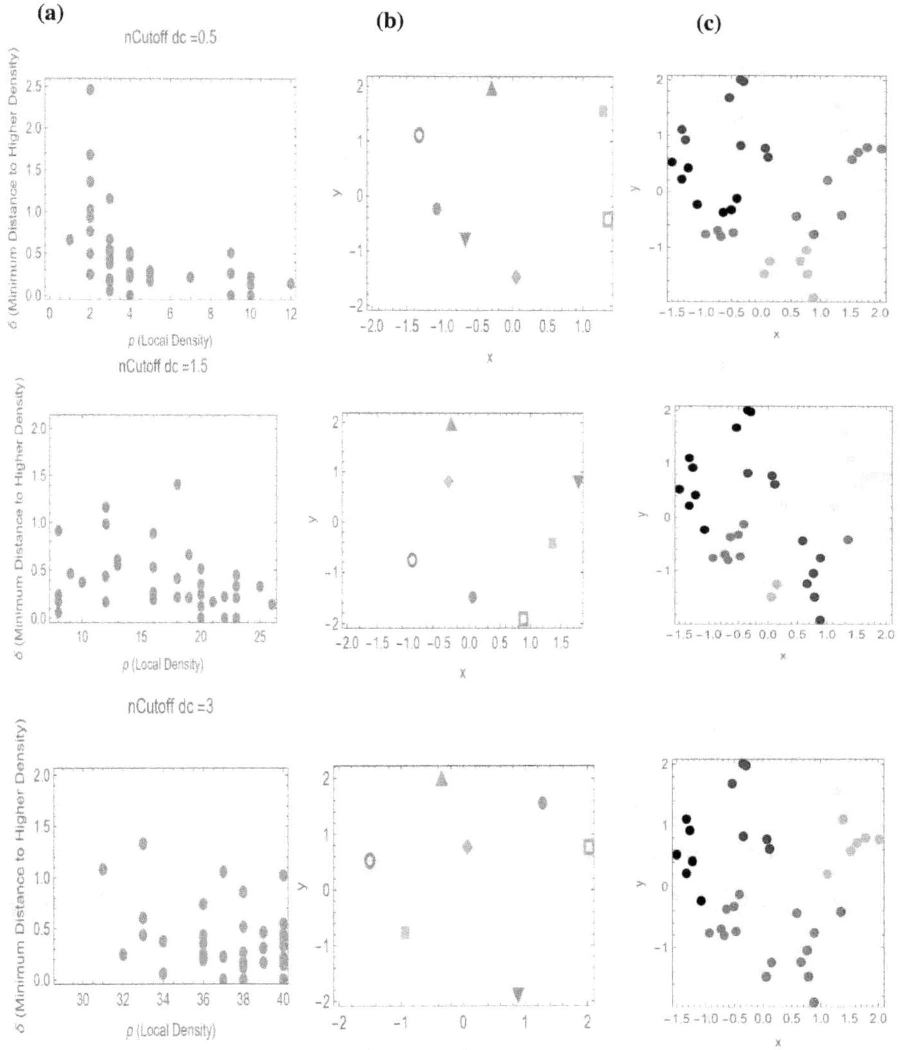

Figure 9: Application of the DPC algorithm to the Teucrium species dataset for different dc values. (a) Decision graph (b) Selected as cluster centers (c) The clustering results

The findings of the computer simulation indicate that the integration of factor analysis and K-means clustering, utilizing the silhouette index, produces successful and effective outcomes for the classification of Teucrium species. Subsequently, utilizing the silhouette coefficients method, we determined that six distinct optimal clusters could be estimated (cf. Fig 8). We proceeded to classify Teucrium species according to their micro morphological characteristics by plotting these clusters in two and three dimensions.

The present study has revealed the potential of employing a combined approach of factor analysis and cluster analysis to elucidate the micromorphological characteristics of Teucrium species. Simulation results demonstrate that combining cluster analysis and factor analysis based on silhouette index results in the creation of a robust tool that can be utilized to identify relationship among numerous variables and a limited number of underlying factors and as well as to categorize samples within extensive datasets.

In order to perform our analysis, the observation data should consist of numerical values obtained according to measurement values that are as accurate as possible. In instances where categorical data are to be utilized, it is imperative that they be converted to numerical values. In the present analysis, the data matrix was subjected to digitization through the implementation of a coding system in which each of the 21 micromorphological characters of the Teucrium species was assigned a numerical value of 1 or 0, respectively, according to the presence or absence of each character (cf. Tables 1 and 2).

The proposed method facilitates the efficient reduction of dimensionality and the clustering of datasets comprising a large number of observations and variables [55]. The performance of the system may not meet the expected standards when dealing with low-dimensional data. It is particularly recommended for multivariate data with large sample sizes. In general, it is expected that using larger samples will produce more accurate results. The significance of our proposed method is amplified when the dataset exhibits a multidimensional structure. For example, there can be more than five variables for each observation. In this study we consider 21 variables for each of the 40 Teucrium species. Therefore, dimension of the data matrix is 40 by 21(cf. Table 2). Furthermore, the application of the proposed method to a data matrix is optimal when the KMO test value exceeds 0.5. The KMO value of the data matrix examined in this study is 0.619, indicating its suitability for the proposed analysis.

It is of paramount importance that the variables exhibit a moderate to high degree of correlation with each other. In general, the correlation between variables should fall within the range of $0.1 \le |R_{ij}| \le 0.9$. As illustrated in Fig.1, the correlation values of the variables with each other are, in general, greater than 10%. This finding lends support to the notion that factor analysis is a suitable method for the analysis of the data. In the analysis proposed in this study, the existence of a correlation between the variables in question is demonstrated, thereby achieving a significant reduction in the size of the method. It is evident that a variable exhibiting no correlation with other variables is inherently incapable of playing an effective role in factor identification. Therefore, it is imperative to regard such a variable as a redundant variable.

It can be stated that our analysis is subject to certain limitations. In the proposed method, observed variables are modeled as linear functions of factors (cf. Eq.1). Nonlinearities are not taken into consideration [55]. The methodology employed to determine the number of factors to be selected to explain m variables in the dataset may yield divergent results. For instance, the Kaiser Criterion [30] and the Scree Plot method (cf. Fig. 4) may yield divergent results [56]. It is generally accepted that determining fewer than two factors will result in a substantial loss of information. Therefore, it is meticulous to determine the number of factors to be at least two.

According to the proposed methodology, the factors with the highest total variance are selected based on the largest eigenvalues for the purpose of dimension reduction. It must be acknowledged that other factors have been disregarded. This may have a negligible effect on the clustering. An excessively high number of factors can result in the model's excessive complexity, while, conversely, an insufficient number can result in the omission of crucial information. In the present analysis, the two factors with the maximum total variance are selected, given that the eigenvalues $\lambda_1 = 5.242$ and $\lambda_2 = 4.196$ are larger than the other eigenvalues (cf. Tables 3 and 5).

The proposed methodology utilizes the silhouette index, which ranges from (-1) to (+1), to ascertain the optimal number of clusters (cf. Table 7). The proximity of this value to zero or its negative value can result in suboptimal cluster formation. In the context of high-dimensional data, the reliability of the silhouette coefficient can be compromised. Given that points in high-dimensional space are generally equidistant from each other, the underlying Euclidean distance of the silhouette coefficient may become meaningless. Therefore, the efficacy of the silhouette coefficient is contingent upon the selection of a distance metric. Different distance metrics can lead to different silhouette scores.

Despite its computational efficiency and ease of implementation, the K-Means clustering algorithm, which is part of the analysis proposed in this book, is subject to certain limitations. These limitations are predicated on certain assumptions, such as the assumption that all features contribute equally to the clustering process and that clusters have similar variances and approximately equal sizes in all dimensions. These assumptions can result in sub-optimal outcomes, including a diminished capacity of the algorithm to differentiate between instances, particularly in high-dimensional data sets comprising features with varying degrees of fitness or when substantial noise is present. Additionally, K-Means may encounter difficulties in identifying natural groupings if clusters exhibit different densities or unequal sizes.

Our analysis is dependent on the number of clusters, k, and should be determined in advance. The utilisation of an erroneous k value has the potential to result in the generation of meaningless clusters. Consequently, the reliability of the clustering index that determines k is of paramount importance (cf. Tables 6 and 7). The proposed methodology yields significant results under the condition that the clusters are distinctly delineated. However, it is important to note that if the clusters are in close proximity to each other, this may potentially result in erroneous outcomes.

In conclusion, we propose that this study constitutes a notable advancement in the investigation of the efficacy of clustering with factors in contrast to the original variables. The versatility and universality of clustering with factor analysis based on silhouette index, combined with its simplicity, may appeal to researchers in different scientific fields. In future research, we intend to employ principal component analysis and factor analysis in combination with various clustering techniques and non-Euclidean distance functions on real data sets, with a view to evaluating their comparative efficacy in cluster analysis. The findings of this study are of significant value to the pharmaceutical industry, particularly in the context of drug discovery, development, production, and marketing.

REFERENCES

1. Stahl, F., Gaber, M.M, Bramer, M. (2023). *Scaling up data mining techniques to large datasets using parallel and distributed processing*. Business Intelligence and Performance Management, Springer, London, pp.243-259.
2. Kokate, U., Deshpande, A., Mahalle, P., Patil, P.(2018). *Data stream clustering techniques, applications, and models: Comparative analysis and discussion*. Big Data Cogn. Comput 2 :32.
3. Kandula,A.R., Sathya, R., Narayana, S. (2023). *Multivariate analysis on personalized cancer data using a hybrid classification model using voting classifier*. International Journal of Intelligent Systems and Applications in Engineering (11) 354–362.
4. Ezugwu, A.E., Ikotun, A.M., Olaide, O., Abualigah, L., Agushaka, O.J., Eke, C.I., Akinyelu, A.(2022). *A comprehensive survey of clustering algorithms: State-of-the-art machine learning applications, taxonomy, challenges, and future research prospects*. Engineering Applications of Artificial Intelligence 110. 104743.
5. Elaziz, M.A., Al-qaness, M.A., Zaid, E.O.A., Lu, S., Ibrahim, R.A., Ewees, A.A. (2021). *Automatic clustering method to segment COVID-19 CT images*, PLoS ONE 16 : e0244416.
6. Dafir, Z., Lamari,Y., Slaoui, S.C. (2021). *A survey on parallel clustering algorithms for big data*. Artificial Intelligence Review (54) 2411-2443.
7. Engelhardt, B.E., Stephens, M.(2010). *Analysis of population structure: A unifying framework and novel methods based on sparse factor analysis*, PLoS Genetics. (6) 1-12.
8. Garson, G.D.(2022). *Factor analysis and dimension reduction in R: A social scientist's toolkit*. First Edition. London.
9. Morin, A. J. S., Myers, N. D., Lee, S.(2020). *Modern factor analytic techniques: Bifactor models, exploratory structural equation modeling (ESEM) and Bifactor-ESEM*. Handbook of Sport Psychology, Fourth Edition, John Wiley & Sons.

10. Bendif, H., Miri, Y.B., Souilah, S., Amina, B., Miara, M.D. (2021). *Phytochemical constituents of Lamiaceae family*. Green and Applied Chemistry (11) 71-88.
11. Genç, G.E., Özcan T, Dirmenci, T.(2017). *Nutlet and leaf micromorphology in some Turkish species of Teucrium L. (Lamiaceae)*, Phytotaxa (312) 71-82.
12. Kremer, D., Stabentheiner, E., Grubesic, R.J., Oberlander, A., Knezevic, S.V., Kozalec, I., Balian, D.(2012). *A morphological and chemotaxonomic study of Teucrium arduini L.in Croatia, and Bosnia and Herzegovina*. Plant Biosystems. (146) 402– 412.
13. Dönmez, A.A.(2006). *Teucrium chasmophyticum Rech. f. (Lamiaceae): A new record for the flora of Turkey*. Turkish Journal of Botany (30) 317-320.
14. Dinç, M., Doğu, S.(2012). *Anatomical and micromorphological studies on Teucrium sect. Isotriodon (Lamiaceae) in Turkey with a taxonomic note*. Biologia (67) 663–672.
15. Aksoy, A., Özcan, T., Girişken, H., Çelik, J., Dirmenci, T.(2020). *A phylogenetic analysis and biogeographical distribution of Teucrium sect. Teucrium (Lamiaceae) and taxonomic notes for a new species from southwest Turkey*. Turkish Journal of Botany (44) 322-337.
16. Çeçen, Ö., Özcan, T. (2021). *A new Teucrium L. (Lamiaceae) species from South Anatolia (Turkey)*. Turkish Journal of Botany (45) 353-370.
17. Kline, P. (1994). *An easy guide to factor analysis*. London; New York: Routledge, 194p.
18. Cudeck, R., MacCallum, R.C. (2007). *Factor Analysis at 100: Historical Developments and Future Directions*. 1st ed. Routledge.
19. Kaiser, H.F.(1974). *An index of factorial simplicity*.Psychometrika(39)31-36.
20. Kaiser, H.F., Derflinger, G. (1990). *Some contrasts between maximum likelihood factor analysis and alpha factor analysis*. Applied Psychological Measurement (14) 29-32.
21. Watkins, M.W. (2021). *A step-by-step guide to exploratory factor analysis with R and R studio.* Taylor & Francis, New York, Routledge Publication.

22. Sneath, P.H., Sokal, R.R. (1973). *Numerical taxonomy: The principles and practice of numerical classification*. W. H. Freeman and Company, San Francisco.

23. Ikotun, A. M., Ezugwu, A.E., Abualigah, L., Abuhaija, B., Heming, J. (2023). *K-means clustering algorithms: A comprehensive review, variants analysis, and advances in the era of big data*. Information Sciences (622) 178-210.

24. Kaufman, L., Rousseeuw, P.J. (1990) *Finding groups in data: An introduction to cluster analysis.* John Wiley and Sons, London: 1990.

25. Yang, J., Lee, J.Y., Choi,M., Joo,Y.(2020). *A new approach to determine the optimal number of clusters based on the gap statistic*. 2nd International Conference on Machine Learning for Networking (MLN), 227–239.

26. Wang, F., Franco-Penya, H. H., Kelleher, J.D., Pugh, J., Ross, R. (2017). *An analysis of the application of simplified silhouette to the evaluation of k-means clustering validity*. 3th International Conference on Machine Learning and Data Mining MLDM, New York, USA.

27. Starczewski, A., Krzyżak, A. (2015). *Performance evaluation of the silhouette index artificial intelligence and soft computing*. Volume 9120 of the series Lecture Notes in Computer Science, Springer International Publishing, 49-58.

28. Bagirov, A.M., Aliguliyev, R.M., Sultanova, N.(2023). *Finding compact and well- separated clusters: Clustering using silhouette coefficients.* Pattern Recognition (135) 109- 144.

29. Tibshirani, R., Walther, G., Hastie, T. (2001). *Estimating the number of clusters in a data set via the gap statistic*. Journal of the Royal Statistical Society, Series B, (63) 411–423.

30. Catell, R.B. (1966). *The scree test for the number of factors*, Multiv.Behav. Res.(1) 245-276.

31. Vandna, K., Yadav, M.P., Kumar, M.(2024). *Elbow method for optimal customer segmentation using K-means clustering*. International Journal of Scientific Research and Engineering Development 7 (3) 842-844.

32. Gordon, D. (1981). *Classification: Methods for the exploratory analysis of multivariate data*. London: Chapman and Hall/CRC.

33. Sharma, S. (1996). *Applied multivariate techniques*. New York, NY, USA: John Wiley & Sons, Inc.

34. Caliński, T., Harabasz, J. (1974). *A dendrite method for cluster analysis*. Communications in Statistics 3(1)1-27.

35. Davies, D. L. , Bouldin, D. W. (1979). *A cluster separation measure*. IEEE Transactions on Pattern Analysis and Machine Intelligence (2) 224-227.

36. Dunn, J. C. (1974). *Well separated clusters and optimal fuzzy partitions*. J. Cybern 4(1) 95-104.

37. Zhou, S., Liu, F., Song, W. (2021). *Estimating the optimal number of clusters via internal validity index*. Neural Process Lett. (53)1013–34.

38. Cevri, M.(2024). *Mathematical analysis of the clustering of Ostracoda concerning their habitat preferences*. Wseas Transactions on Systems (23) 314-321.

39. El-Mandouh, A.M., Abd-Elmegid, L.A., Mahmoud, H.A., Haggag, M.H.(2019). *Optimized K-means clustering model based on gap statistic*. International Journal of Advanced Computer Science and Applications (10) 183-188.

40. Karim, I. K. , Daud, H. B., Zainuddin, N.B., Sokkalingam, R., Farooq, M., Baig, M. E., Ayub, G., Zafar, M.(2024). *Determining the optimal number of clusters by enhanced gap statistic in K- mean algorithm*. Egyptian Informatics Journal (27)100504.

41. Xu, R., Xu, J., Wunsch, D. C.(2012). *A comparison study of validity indices on swarm intelligence based clustering*. IEEE Transactions on Systems, Man, and Cybernetics, Part B (Cybernetics) 42(4)1243-1256.

42. Senthilkumar, A., Metilda, M.(2019). *An extensive investigation of outlier detection by cluster validation indices*. Ciencia e Tecnica Vitivinicola-A Science and Technology Journal, 34 (2) 22-32.

43. Öztürk, F.E, Demirel, N.(2023). *Comparison of the methods to determine optimal number of clusters*. Veri Bilim Derg. 6(1) 34-45.

44. Kaufman, L., Rousseeeuw, P.(1987). *Clustering by means of medoids*. Faculty of Mathematics and Informatics.

45. Jain, A. K. , Dubes, R. C. (1988). *Algorithms for clustering data*. Upper Saddle River, NJ, USA: Prentice-Hall, Inc.

46. Ester, M., Kriegel, H.P., Sander, J., Xu, X.(1996). *A density-based algorithm for discovering clusters in large spatial databases with noise*. Second International Conference on Knowledge Discovery and Data Mining (KDD'96), 226–231.

47. Jarvis, R. A, Patrick, E. A.(1973). *Clustering using a similarity measure based on shared near neighbors*. Computers, IEEE Transactions on, vol. C-22(11)1025-1034.

48. Grygorash, O., Zhou, Y., Jorgensen, Z.(2006). *Minimum spanning tree-based clustering algorithms*. Proceedings of the 18th IEEE International Conference on Tools with Artificial Intelligence (ICTAI '06) 73–81.

49. Jia, H., Ding, S., Xu, X., Nie, R.(2014). *The latest research progress on spectral clustering*. Neural Compute. Applicant. 24 (7) 1477–1486.

50. Rodriguez, A., Liao, A.(2014). *Clustering by fast search and find of density peaks*. Science 344(6191)1492-1496.

51. Huang, J.Z., Ng, M.K., Rong, H., Li, Z.(2005). *Automated variable weighting in k-means type clustering*. IEEE Transactions on Pattern Analysis and Machine Intelligence 27 (5) 657-668.

52. Scholz, M., Wimmer, T.(2021). *A comparison of classification methods across different data complexity scenarios and datasets*. Expert Systems with Applications (168)114217.

53. Cevri, M. (2025). An efficient algorithm based on cluster analysis for exploring structure of large multivariate datasets. Computers in Biology and Medicine, Vol. 197, 111016.

54. Wei, X., Peng, M., Huang, H., Zhou, Y.(2023). *An overview on density peaks clustering*. Neurocomputing (554) 126633.

55. Fabrigar, L.R., Wegener, D.T.(2012). *Exploratory factor analysis*. Oxford University Press. New York.

56. Finch, W. H. (2023). *A comparison of methods for determining the number of factors to retain with exploratory factor analysis of dichotomous data*. Psych 5(3) 1004-1018.

Publisher: Eliva Press Global Ltd

Email: info@elivapress.com

www.ingramcontent.com/pod-product-compliance
Lightning Source LLC
LaVergne TN
LVHW022321200926
848481LV00004B/1388
* 9 7 8 9 9 9 9 3 4 5 8 8 0 *